Leonard S Vincent

The Carolina Locust (*Dissosteira carolina*), a common American grasshopper. (Enlarged about 2½ times)

INSECTS

THEIR WAYS AND MEANS
OF LIVING

By

Robert Evans Snodgrass

DOVER PUBLICATIONS, INC.
NEW YORK

Published in Canada by General Publishing Company, Ltd., 30 Lesmill Road, Don Mills, Toronto, Ontario.

Published in the United Kingdom by Constable and Company, Ltd., 10 Orange Street, London WC 2.

This Dover edition, first published in 1967, is an unabridged republication of the work originally published by the Smithsonian Institution Series, Inc., in 1930.

Twelve illustrations that appeared in full color in the original edition are here reproduced in black and white.

International Standard Book Number: 0-486-21801-5
Library of Congress Catalog Card Number: 67-17986

Manufactured in the United States of America
Dover Publications, Inc.
180 Varick Street
New York, N. Y. 10014

CONTENTS

ILLUSTRATIONS

LIST OF PLATES

LIST OF TEXT FIGURES

INSECTS
THEIR WAYS AND MEANS
OF LIVING

PREFACE

In the early days of zoology there were naturalists who spent much time out of doors observing the ways of the birds, the insects, and the other creatures of the fields and woods. These men were not steeped in technical learning. Nature was a source of inspiration and a delight to them; her manifestations were to be taken for granted and not questioned too closely. A mind able to accept appearances for truth can express itself in the words of everyday language—for language was invented long ago when people did not bother themselves much with facts—and some of those early writers, inspired direct from nature, have left us a delightful literature based on their observations and reflections on the things of nature. The public has liked to read the works of these men because they tell of interesting things in an interesting way and in words that can be understood.

At the same time there was another class of nature students who did not care particularly what an animal did, but who wanted to know how it was made. The devotees of this cult looked at things through microscopes; they dissected all kinds of creatures in order to learn their construction and their structural relationships. But they found many things on the inside of animals that had never been named, so for these things they invented names; and when their books were printed the public could not read them because of the strange words they contained. Moreover, since nature does not usually embellish her hidden works, the anatomists could not enhance their writings with descriptive metaphors in the way the outdoor naturalists could. Consequently, the students of structure have never come into favor with the reading public, and their works are denounced as dry and tedious.

PREFACE

Then there arose still another group of inquiring minds. Members of this school could not see anything worth while in knowing merely either what an animal did or how it was made. They devoted their efforts to discovering the secrets of its workings. They invented instruments for measuring the power of its muscles, for testing the nature of the force that resides in its nerves; they made analyses of its food and its tissues; they devised all kinds of experiments for revealing the causes of its behavior. The workers in this branch, the physiologists, had to have a considerable grounding in physics and chemistry; consequently they came to write more or less in the languages of those sciences and to express themselves in chemical and mathematical formulae. Their writings are hard for the public to understand. Their statements, moreover, are often at odds with preconceived ideas, since preconceived ideas are conceived in ignorance, and the public at large does not take to this sort of thing—it cherishes above all its inherited opinions.

Therefore the old-time naturalist is still venerated, as he deserves to be, and those who call themselves "nature lovers" still like to decry the laboratory worker as an evil being who would take the beauty from nature and destroy the soul of man. A modern writer of the old school may sell his wares, but when something goes wrong with his stomach or his nerves, or when his plants or his animals are attacked by disease, it is the knowledge of the laboratory scientist that comes to his aid.

The reason that the specific truths of nature must be found out in laboratories is that there are too many things mixed together in the fields. The laboratory naturalist endeavors to untangle the confusion of elements in the outdoor environment and to isolate the different factors that affect the life and behavior of an animal, in order that he may be sure with just what he is dealing in his efforts to determine the value of each one separately. By creating a set of artificial environments in each of which

[ii]

only one natural factor is allowed to be operative at the same time, he is in a position to observe correctly, after repeated experiments, just what effects proceed from this cause and what from that.

Nature study, in the superficial sense, may be entertaining. We of the present age, however, must learn to take a deeper insight into the lives of the other living things about us. Insects, for example, are not curiosities; they are creatures in common with ourselves bound by the laws of the physical universe, which laws decree that everything alive must live by observing the same elemental principles that make life possible. It is only in the ways and means by which we comply with the conditions laid down by physical nature that we differ.

Many sincere people find it difficult to believe in evolution. Their difficulty arises largely from the fact that they look to the differences in structure between the diverse types of living things and do not see the unity in function that underlies all physical forms of life. Consequently they do not understand that evolution means the progressive structural *divergence* of the various life forms from one another, resulting from the different ways that each has adopted and perfected for accomplishing the same ends. Man and the insects represent the extremities of two most divergent lines of animal evolution, and by reason of the very disparity in structure between us the bond of unity in function becomes all the more apparent. A study of insects, therefore, will help us the better to understand ourselves in so far as it helps us to grasp the fundamental principles of life.

Some writers seem to think that the sole purpose of writing is that it shall be read. Just as reasonable would it be to claim that the only purpose of food is that it shall be eaten. In the following chapters the reader is offered an entomological menu in which the consideration of nutrient value and the requirements of a balanced meal have been given first attention. As a concession to

PREFACE

palatability, however, as much as possible of the distasteful matter of technical terminology has been extracted, and an attempt has been made to avoid the pure scientific style of literary cuisine, which forbids the use of all those ingredients whose object is that of inflation but which, if properly admixed, will greatly aid in the process of digestion.

Much of the material in several chapters is taken from articles already printed in the *Annual Reports* of the Smithsonian Institution. The original drawings of most of the color plates and line cuts are the property of the United States Bureau of Entomology, though some of them are here published for the first time.

R. E. S.

INSECTS

THEIR WAYS AND MEANS OF LIVING

CHAPTER I

THE GRASSHOPPER

SOMETIME in spring, earlier or later according to the latitude or the season, the fields, the lawns, the gardens, suddenly are teeming with young grasshoppers. Comical little fellows are they, with big heads, no wings, and strong hind legs (Fig. 1). They feed on the fresh herbage and hop lightly here and there, as if their existence in no way involved the mystery of life nor raised any questions as to why they are here, how they came to be here, and whence they came. Of these questions, the last is the only one to which at present we can give a definite answer.

If we should search the ground closely at this season, it might be possible to see that the infant and apparently motherless grasshoppers are delivered into the visible world from the earth itself. With this information, a nature student of ancient times would have been satisfied —grasshoppers, he would then announce, are bred spontaneously from matter in the earth; the public would believe him, and thereafter would countenance no contrary opinion. There came a time in history, however, when some naturalist succeeded in overthrowing this idea and established in its place the dictum that every life comes from an egg. This being still our creed, we must look for the grasshopper's egg.

INSECTS

The entomologist who plans to investigate the lives of grasshoppers finds it easier to begin his studies the year before; instead of sifting the earth to find the eggs from which the young insects are hatched in the spring, he observes the mature insects in the fall and secures a supply of eggs freshly laid by the females, either in the field or in cages properly equipped for them. In the laboratory then

Fig. 1. Young grasshoppers

he can closely watch the hatching and observe with accuracy the details of the emergence. So, let us reverse the calendar and take note of what the mature grasshoppers of last season's crop are doing in August and September.

First, however, it is necessary to know just what insect is a grasshopper, or what insect we designate by the name; for, unfortunately, names do not always signify the same thing in different countries, nor is the same name always applied to the same thing in different parts of the same country. It happens to be thus with the term "grasshopper." In most other countries they call grasshoppers "locusts," or rather, the truth is that we in the United States call locusts "grasshoppers," for we must, of course, concede priority to Old World usage. When you read of a "plague of locusts," therefore, you must understand "grasshoppers." But a swarm of "seventeen-year locusts" means quite another insect, neither locust nor grasshopper —correctly, a cicada. All this mix-up of names and many other misfits in our popular natural history parlance we can blame probably on the early settlers of our States, who bestowed upon the creatures encountered in the New World the names of animals familiar at home; but, having no zoologists along for their guidance, they made many errors of identification. Scientists have sought to establish a better state of nomenclatural affairs by creating a set of international names for all living things, but since their names are in Latin, or Latinized Greek, they are seldom practicable for everyday purposes.

Knowing now that a grasshopper is a locust, it only needs to be said that a true locust is any grasshopperlike insect with short horns, or *antennae* (see Frontispiece). A similar insect with long slender antennae is either a katydid (Figs. 23, 24), or a member of the cricket family (Fig. 39). If you will collect and examine a few specimens of locusts, which we will proceed to call grasshoppers, you may observe that some have

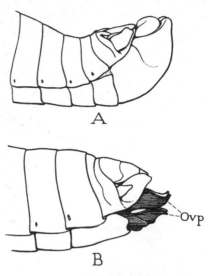

Fig. 2.   The end of the body of a male and a female grasshopper

The body, or abdomen, of a male (A) is bluntly rounded; that of the female (B) bears two pairs of thick prongs, which constitute the egg-laying organ, or ovipositor (*Ovp*)

the rear end of the body smoothly rounded and that others have the body ending in four horny prongs. The second kind are females (Fig. 2 B); the others (A) are males and may be disregarded for the present. It is one of the provisions of nature that whatever any creature is compelled by its instinct to do, for the doing of that thing it is provided with appropriate tools. Its tools, however, unless it is a human animal, are always parts of its body, or of its jaws or its legs. The set of prongs at the end of the body of the female grasshopper constitutes a digging tool, an instrument by means of which the insect makes a hole in the ground wherein she deposits her eggs. Entomologists call the organ an *ovipositor*, or egg-placer. Figure 2 B

FIG. 3. A female grasshopper in the position of depositing a pod of eggs in a hole in the ground dug with her ovipositor. (Drawn from a photograph in U. S. Bur. Ent.)

shows the general form of a grasshopper's ovipositor; the prongs are short and thick, the points of the upper pair are curved upward, those of the lower bent downward.

When the female grasshopper is ready to deposit a batch of eggs, she selects a suitable spot, which is almost any place in an open sunny field where her ovipositor can penetrate the soil, and there she inserts the tip of her organ with the prongs tightly closed. When the latter are well within the ground, they are probably spread apart so as to compress the earth outward, for the drilling process brings no detritus to the surface, and gradually the end of the insect's body sinks deeper and deeper, until a considerable length of it is buried in the ground (Fig. 3).

Now all is ready for the discharge of the eggs. The exit duct from the tubes of the ovary, which are filled with eggs already ripe, opens just below and between the bases of the lower prongs of the ovipositor, so that, when the upper and lower prongs are separated, the eggs escape from the passage between them. While the eggs are being placed in the bottom of the well, a frothy gluelike substance from the body of the insect is discharged over them. This sub- stance hardens about the eggs as it dries, but not in a solid mass, for its frothy nature leaves it full of cavities, like a sponge, and affords the eggs, and the young grasshoppers when they hatch, an abundance of space for air. To the outside of the covering substance, while it is fresh and sticky, particles of earth adhere and make a finely granular coating

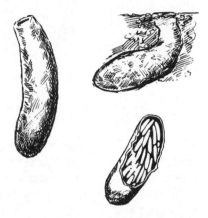

FIG. 4. Egg pods of a grasshopper, show- ing various shapes; one opened exposing the eggs within. (Much enlarged)

over the mass, which, when hardened, looks like a small pod or capsule that has been molded into the shape of the cavity containing it (Fig. 4). The number of eggs within each pod varies greatly, some pods containing only half a dozen eggs, and others as many as one hundred and fifty. Each female also deposits several batches of eggs, each lot in a separate burrow and pod, before her egg supply is exhausted. Some species arrange the eggs regularly in the pods, while others cram them in hap- hazard.

[5]

The egg of a grasshopper is elongate-oval in shape (Fig. 5), those of ordinary-sized grasshoppers being about three-sixteenths of an inch in length, or a little longer.

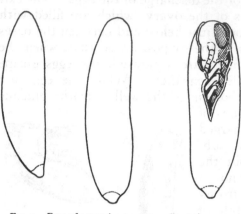

The ends of the eggs are rounded or bluntly pointed, and the lower extremity (the egg being generally placed on end) appears to have a small cap over it. One side of the egg is always more curved than the opposite side, which may be almost straight.

FIG. 5.   Eggs of a grasshopper; one split at the upper end, showing the young grasshopper about to emerge

The surface is smooth and lustrous to the naked eye, but under the microscope it is seen to be marked off by slightly raised lines into many small polygonal areas.

Within each egg is the germ that is to produce a new grasshopper. This germ, the living matter of the egg, is but a minute fraction of the entire egg contents, for the bulk of the latter consists of a nutrient substance, called yolk, the purpose of which is to nourish the embryo as it develops. The tiny germ contains in some form, that even the strongest microscope will not reveal, the properties which will determine every detail of structure in the future grasshopper, except such as may be caused by external circumstances. It would be highly interesting to follow the course of the development of the embryo insect within the egg, and most of the important facts about it are known; but the story would be entirely too long to be given here, though a few things about the grasshopper's development should be noted.

THE GRASSHOPPER

The egg germ begins its development as soon as the eggs are laid in the fall. In temperate or northern latitudes, however, low temperatures soon intervene, and development is thereby checked until the return of warmth in the spring—or until some entomologist takes the eggs into an artificially heated laboratory. The eggs of some species of grasshoppers, if brought indoors before the advent of freezing weather and kept in a warm place, will proceed with their development, and young grasshoppers will emerge from them in about six weeks. On the other hand, the eggs of certain species, when thus treated, will not hatch at all; the embryos within them reach a certain stage of development and there they stop, and most of them never will resume their growth unless they are subjected to a freezing temperature! But, after a thorough chilling, the young grasshoppers will come out, even in January, if the eggs are then transferred to a warm place.

To refuse to complete its development until frozen and then warmed seems like a preposterous bit of inconsistency on the part of an insect embryo; but the embryos of many kinds of insects besides the grasshopper have this same habit from which they will not depart, and so we must conclude that it is not a whim but a useful physiological property with which they are endowed. The special deity of nature delegated to look after living creatures knows well that Boreas sometimes oversleeps and that an egg laid in the fall, if it depended entirely on warmth for its development, might hatch that same season if mild weather should continue. And then, what chance would the poor fledgling have when a delayed winter comes upon it? None at all, of course, and the whole scheme for perpetuation of the species would be upset. But, if it is so arranged that development within the egg can reach completion only after the chilling effect of freezing weather, the emergence of the young insect will be deferred until the return of warmth in the spring, and thus the species will have a guarantee that its members will not be cut down by unseasonable hatching. There are, however, species not thus insured, and these do suffer losses from fall hatching every time winter makes a late arrival. Eggs laid in the spring are designed to hatch the same season, and the eggs of species that live in warm climates never require freezing for their development.

The tough shell of the grasshopper's egg is composed of two distinct coats, an outer, thicker, opaque one of a pale brown color, and an inner one which is thin and transparent. Just before hatching, the outer coat splits open in an irregular break over the upper end of the egg, and usually half or two-thirds of the way down the flat side. This outer coat can easily be removed artificially, and the inner coat then appears as a glistening capsule, through the semitransparent walls of which the little grasshopper inside can be seen, its members all tightly folded beneath its body. When the hatching takes place normally, however, both layers of the eggshell are split, and the young grasshopper emerges by slowly making its way out of the cleft (Fig. 6).

FIG. 6. Young grasshopper emerging from its eggshell

Newly-hatched grasshoppers that have come out of eggs which some meddlesome investigator has removed from their pods for observation very soon proceed to shed an outer skin from their bodies. This skin, which is already loosened at the time of hatching, appears now as a rather tightly fitting garment that cramps the soft legs and feet of the delicate creature within it. The latter, however, after a few forward heaves of the body, accompanied by expansions of two swellings on the back of the neck (Fig. 6), succeeds in splitting the skin over the neck and the back of the head, and the pellicle then rapidly shrinks and slides down over the body. The insect, thus first exposed, liberates itself from the shriveled remnant of its hatching skin, and becomes a free new creature in the world. Being a grasshopper, it proceeds to jump, and with its first efforts clears a distance of four or five inches, something like fifteen or twenty times the length of its own body.

When the young locusts hatch under normal undisturbed conditions, however, we must picture them as coming out of the eggs into the cavernous spaces of the egg pod, and all buried in the earth. They are by no means yet free creatures, and they can gain their liberty only by burrowing upward until they come out at the surface of the ground. Of course, they are not very far beneath the surface, and most of the way will be through the easily penetrated walls of the cells of the egg covering. But above the latter is a thin layer of soil which may be hard-packed after the winter's rains, and breaking through this layer can not ordinarily be an easy task. Not many entomologists have closely watched the newly-hatched grasshopper emerge from the earth, but Fabre has studied them under artificial conditions, covered with soil in a glass tube. He tells of the arduous efforts the tiny creatures make, pressing their delicate bodies upward through the earth by means of their straightened hind legs, while the vesicles on the back of the neck alternately contract and expand to widen the passage above. All this, Fabre says, is done before the hatching skin is shed, and it is only after the surface is reached and the insect has attained the freedom of the upper world that the inclosing membrane is cast off and the limbs are unencumbered.

The things that insects do and the ways in which they do them are always interesting as mere facts, but how much wiser might we be if we could discover why they do them! Consider the young locust buried in the earth, for example, scarcely yet more than an embryo. How does it know that it is not destined to live here in this dark cavity in which it first finds itself? What force activates the mechanism that propels it through the earth? And finally, what tells the creature that liberty is to be found above, and not horizontally or downward? Many people believe that these questions are not to be answered by human knowledge, but the scientist has faith in the ultimate solution of all problems, at least in terms of the elemental forces that control the activities of the universe.

We know that all the activities of animals depend upon the nervous system, within which a form of energy resides that is delicately responsive to external influences. Any kind of energy harnessed to a physical mechanism will produce results depending on the construction of the mechanism. So the effects of the nerve force within a living animal are determined by the physical structure of the animal. An instinctive action, then, is the expression of nerve energy working in a particular kind of machine. It would involve a digression too long to explain here the modern conception of the nature of instinct; it is sufficient to say that something in the surroundings encountered by the newly-hatched grasshopper, or some substance generated within it, sets its nerve energy into action, that the nerve energy working on a definite mechanism produces the motions of the insect, and that the mechanism is of such a nature that it works against the pull of gravity. Hence the creature, if normal and healthy in all respects, and if the obstacles are not too great, arrives at the surface of the ground as inevitably as a submerged cork comes to the surface of the water. Some readers will object that an idea like this destroys the romance of life, but whoever wants romance must go to the fiction writers; and even romance is not good fiction

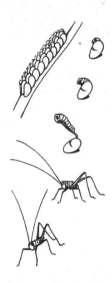

FIG. 7. Eggs of a species of katydid attached to a twig; the young insect in successive stages of emerging from an egg; and the newly-hatched young

unless it represents an effort to portray some truth.

Insects hatched from eggs laid in the open may begin life under conditions a little easier than those imposed upon the young grasshopper. Here, for example (Fig. 7), are some eggs of insects belonging to the katydid family. They look like flat oval seeds stuck in overlapping rows, some on a twig, others along the edge of a leaf. When about to hatch, each egg splits halfway down one edge and crosswise on the exposed flat surface, allowing a flap to open on this side, which gives an easy exit to the young insect about to emerge. The latter is inclosed in a delicate transparent sheath, within which its long legs and antennae are closely doubled up beneath the body; but when the egg breaks open, the sheath splits also, and as the young insect emerges it sheds the skin and leaves it within the shell. The new creature has nothing to do now but to stretch its long legs, upon which it walks away, and, if given suitable food, it will soon be contentedly feeding.

Let us now take closer notice of the little grasshoppers (Fig. 8) that have just come into the great world from the dark subterranean chambers of their egg-pods. Such an inordinately large head surely, you would say, must over-balance the short tapering body, though supported on three pairs of legs. But, whatever the proportions, nature's works never have the appearance of being out of drawing; because of some law of recompense, they never give you the uneasy feeling of an error in construction. In spite of its enormous head, the grasshopper infant is an agile creature. Its six legs are all attached to the part of the body immediately behind the head, which is known as the *thorax* (Fig. 63, *Th*), and the rest of the body, called the *abdomen* (*Ab*), projects free without support. An insect, according to its name, is a creature divided into parts, for "insect" means "in-cut." A fly or a wasp, therefore, comes closer to being the ideal insect; but, while not literally in-sected between the thorax and abdomen, the grasshopper, like the fly and the wasp and all other insects, consists of a head, a thorax bearing the legs, and a terminal abdomen (Fig. 63). On the head is located a pair of long, slender *antennae* (*Ant*) and a pair of large eyes (*E*). Winged insects have usually two pairs of wings attached to the back of the thorax (W_2, W_3).

The outside of the insect's body, instead of presenting a continuous surface like that of most animals, shows many encircling rings where the hard integument appears to be infolded, as it really is, dividing each body region except the head into a series of short overlapping sections. These

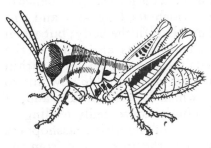

body sections are called *segments*, and all insects and their relatives, including the centipedes, the shrimps, lobsters, and crabs, and the scorpions and spiders, are segmented animals. The insect's thorax consists of three segments, the first of which carries the first pair of legs, the second

FIG. 8. A young grasshopper, or nymph, in the second stage after hatching

the middle pair of legs, and the third the hind pair of legs. The abdomen usually consists of ten or eleven segments, but generally has no appendages, except a pair of small peglike organs at the end known as the *cerci*, and, in the adult female, the prongs of the ovipositor (Fig. 2 B), which belong to the eighth and ninth segments.

The head, besides carrying the antennae (Fig. 63, *Ant*), has three pairs of appendages grouped about the mouth, which serve as feeding organs and are known collectively as the *mouth parts*. The presence of four pairs of appendages on the head raises the question, then, as to why the head is not segmented like the thorax and the abdomen. At an early stage of embryonic growth the head *is* segmented, and each pair of its appendages is borne by a single segment, but the head segments are later condensed

[12]

into the solid capsule of the cranium. Thus we see that the entire body of an insect is composed of a series of segments which have become grouped into the three body regions. Note that the insect does not have a "nose" or any breathing apertures on its head. It has, however, many nostrils, called *spiracles* (Fig. 70, *Sp*), distributed along each side of the thorax and the abdomen. Its breathing system is quite different from ours, but will be described in another chapter treating of the internal organization (page 114).

Most young insects grow rapidly because they must compress their entire lives within the limits of a single season. Generally a few weeks suffice for them to reach maturity, or at least the mature growth of the form in which they leave the egg, for, as we shall see, many insects complicate their lives by having several different stages, in each of which they present quite a different form. The grasshopper, however, is an insect that grows by a direct course from its form at hatching to that of the adult, and at all stages it is recognizable as a grasshopper (Fig. 9). A young moth, on the other hand, hatching in the

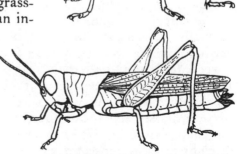

Fig. 9. The metamorphosis of a grasshopper, *Melanoplus atlanus*, showing its six stages of development from the newly-hatched nymph to the fully-winged adult. (Twice natural size)

form of a caterpillar, has no resemblance to its parent, and the same is true of a young fly, which is a maggot, and of the grublike young of a bee. The changes of form that insects undergo during their growth are known as *metamorphosis*. There are different degrees of such transformation; the grasshopper and its relatives have a simple metamorphosis.

An insect differs from a vertebrate animal in that its muscles are attached to its skin. Most species of insects have the skin hardened by the formation of a strong outside *cuticula* to give a firm support to the muscles and to resist their pull. This function of the cuticula, however, imposes a condition of permanency on it after it is once formed. As a consequence the growing insect is confronted with the alternatives, after reaching a certain size, of being cramped to death within its own skin, or of discarding the old covering and getting a new and larger one. It has adopted the course of expediency, and periodically *molts*. Thus it comes about that the life of an insect progresses by stages separated by the molts, or the shedding of the cuticula.

The grasshopper makes six molts between the time of hatching and its attainment of the final adult form, a period of about six weeks, and goes through six postembryonic stages (Fig. 9). The first molt is the shedding of the embryonic skin, which, we have seen, takes place normally as soon as the young insect emerges from the earth. The grasshopper now lives uneventfully for about a week, feeding by preference on young clover leaves, but taking almost any green thing at hand. During this time its abdomen lengthens by the extension of the membranes between its segments, but the hard parts of the body do not change either in size or in shape. At the end of seven or eight days, the insect ceases its activities and remains quiet for a while until the cuticula opens in a lengthwise split over the back of the thorax and on the top of the head. The dead skin is then cast off, or rather, the grasshopper emerges from it, carefully pulling its legs and antennae from their containing sheaths. The whole process consumes only a few minutes. The emerged grasshopper is now entering its third stage after hatching, but the shedding of the hatching skin is usually not counted in the series of molts, and the first subsequent molt, then, we will say, ushers it into its second stage of aboveground life. In this state the insect is different in some respects from what it was in the first stage: it is not only larger, but the body is longer in proportion to the size of the head, as are also the antennae, and particularly the hind legs. Again the insect becomes active and pursues its routine life for another week; then it undergoes a second molting, accompanied by changes in form and proportions that make it a little more like a mature grasshopper. After shedding its cuticula on three succeeding occasions, it appears in the adult form, which it will retain throughout the remainder of its life.

The grasshopper developed its legs, its antennae, and most of its other organs while it was in the egg. It was hatched, however, without wings, and yet, as everyone knows, most full-grown grasshoppers have two pairs of wings (Fig. 63, W_2, W_3), one pair attached to the back of the middle segment of the thorax, the other to the third segment. It has acquired its wings, therefore, during its growth from youth to maturity, and by examining the insect in its different stages (Fig. 9), we may learn something of how the wings are developed. In the first stage, evidence of the coming wings is scarcely apparent, but in the second, the lower hind angles of the plates covering the back of the second and third thoracic segments are a little enlarged and project very slightly as a pair of lobes. In the third stage, the lobes have increased in size and may now be suspected of being rudiments of the wings, which, indeed, they are. At the next molt, when the insect enters its fourth stage, the little wing pads are turned upward and laid over the back, which disposition not only reverses the natural position of the wings, but brings the hind pair outside the front pair. At the next molt, the wings retain their reversed positions, but they are once more increased in size, though they still remain far short of the dimensions of the wings of an adult grasshopper.

At the time of the last molt, the grasshopper takes a position with its head downward on some stem or twig, which it grasps securely with the claws of its feet. Then, when its cuticula splits, it crawls downward out of the skin. Once free, however, it reverses its position, and the wisdom of this act is seen on observing the rapidly expanding and lengthening wings, which can now hang downward and spread out freely without danger of crumpling. In a quarter of an hour the wings have enlarged from small, insignificant pads to long, thin, membranous fans that reach to the tip of the body. This rapid growth is explained by the fact that the wings are hollow sacs; their visible increase in size is a mere distention of their wrinkled walls, for they were fully formed beneath the old cuticula and lay there before the molt as little crumpled wads, which, when released by the removal of the cases that cramped them, rapidly spread out to their full dimensions. Their thin, soft walls then come together, dry, and harden, and the limp, flabby bags are converted into organs of flight.

It is important to understand the process of molting as it takes place in the grasshopper, because the processes of metamorphosis, such as those which accomplish the transformation of a caterpillar into a butterfly, differ only in degree from those that accompany the shedding of the skin between any two stages of the grasshopper's life. The principal growth of the insect is made during those resting periods preceding the molts. It is then that the various parts enlarge and make whatever alterations in shape they are to have. The old cuticula is already loosened and the changes go on beneath it, while at the same time a new cuticula is generated over the remodeled surfaces. The increased size of the antennae, legs, and wings causes them to be compressed in the narrow space between the new and the old cuticula, and, when the latter is cast off, the crumpled appendages expand to their full size. The observer then gets the impression that he is witnessing a sudden transformation. The impression, however, is a false one; what is really going on is comparable with the display of new dresses and coats that the merchant puts into his show windows at the proper season for their use, which he has just unpacked from their cases but which were produced in the factories long before.

The adult grasshoppers lead prosaic lives, but, like a great many good people, they fill the places allotted to them in the world, and see to it that there will be other occupants of their own kind for these same places when they themselves are forced to vacate. If they seldom fly high, it is because it is not the nature of locusts to do so; and if, in the East, one does sometimes soar above his fellows, he accomplishes nothing, unless he happens to land on the upper regions of a Manhattan skyscraper, when he may attain the glory of a newspaper mention of his exploit—most likely, though, with his name spelled wrong.

On the other hand, like all common folk born to obscurity and enduring impotency as individuals, the grasshopper in masses of his kind becomes a formidable creature. Plagues of locusts are of historic renown in countries south of the Mediterranean, and even in our own country hordes of grasshoppers known as the Rocky Mountain locust did such damage at one time in the States of the Middle West that the government sent out a commission of entomologists to investigate them. This was in the years following the Civil War, when, for some reason, the locusts that normally inhabited the Northwest, east of the Rocky Mountains, became dissatisfied with their usual breeding grounds and migrated in great swarms into the States of the Mississippi valley, where they brought destruction to all kinds of crops wherever they chanced to alight. In the new localities they would lay their eggs, and the young of the next season, after acquiring their wings, would migrate back toward the region whence the parent swarm had come the year before.

The entomologists of the investigating commission in the year 1877 tell us that on a favorable day the migrating locusts "rise early in the forenoon, from eight to ten o'clock, and settle down to eat from four to five in the afternoon. The rate at which they travel is variously estimated from three to fifteen or twenty miles an hour, determined by the velocity of the wind. Thus, insects which began to fly in Montana by the middle of July may not reach Missouri until August or early September, a period of about six weeks elapsing before they reach their destined breeding grounds." The appearance of a swarm in the air was described as being like that of "a vast body of fleecy clouds," or a "cloud of snowflakes," the mass of flying insects "often having a depth that reaches from comparatively near the ground to a height that baffles the keenest eye to distinguish the insects in the upper stratum." It was estimated that the locusts could fly at an elevation of two and a half miles from the general surface of the ground, or 15,000 feet above sea level. The descending swarm falls upon the country "like a plague or a blight," said one of the entomologists of the commission, Dr. C. V. Riley, who has left us the following graphic picture of the circumstances:

The farmer plows and plants. He cultivates in hope, watching his growing grain in graceful, wave-like motion wafted to and fro by the warm summer winds. The green begins to golden; the harvest is at hand. Joy lightens his labor as the fruit of past toil is about to be realized. The day breaks with a smiling sun that sends his ripening rays through laden orchards and promising fields. Kine and stock of every sort are sleek with plenty, and all the earth seems glad. The day grows. Suddenly the sun's face is darkened, and clouds obscure the sky. The joy of the morn gives way to ominous fear. The day closes, and ravenous locust-swarms have fallen upon the land. The morrow comes, and, ah! what a change it brings! The fertile land of promise and plenty has become a desolate waste, and old Sol, even at his brightest, shines sadly through an atmosphere alive with myriads of glittering insects.

Even today the farmers of the Middle Western States are often hard put to it to harvest crops, especially alfalfa and grasses, from fields that are teeming with hungry grasshoppers. By two means, principally, they seek relief from the devouring hordes. One method is that of driving across the fields a device known as a "hopperdozer," which collects the insects bodily and destroys them. The dozer consists essentially of a long shallow pan, twelve or fifteen feet in length, set on low runners and provided with a high back made either of metal or of cloth stretched over a wooden frame. The pan contains water with a thin film of kerosene over it. As the dozer is driven over the field, great numbers of the grasshoppers that fly up before it either land directly in the pan or fall into it after striking the back, and the kerosene film on the water does the rest, for kerosene even in very small quantity is fatal to the insects. In this manner, many bushels of dead locusts are taken often from each acre of an alfalfa field; but still great numbers of them escape, and the dozer naturally can not be used on rough or uneven ground, in pastures, or in fields with standing crops. A more generally effective method of killing the pests is that of poisoning them. A mixture is prepared of bran, arsenic, cheap molasses, and water, sufficiently moist to adhere in small lumps, with usually some substance added which is supposed to make the "mash" more attractive to the insects. The deadly bait is then finely broadcast over the infested fields.

While such methods of destruction are effective, they bear the crude and commonplace stamp of human ways. See how the thing is done when insect contends against insect. A fly, not an ordinary fly, but one known to entomologists as *Sarcophaga kellyi* (Fig. 10), being named after Dr. E. O. G. Kelly, who has given us a description of its habits, frequents the fields in Kansas where grasshoppers are abundant. Individuals of this fly, according to Doctor Kelly's account, are often seen to dart after grasshoppers on the wing and strike against them. The stricken insect at once drops to the ground. Examination reveals no physical injury to the victim, but on a close inspection there may be found adhering to the under surface of a

FIG. 10. A fly whose larvae are parasitic on grasshoppers, *Sarcophaga kellyi*. (Much enlarged)

wing several tiny, soft, white bodies. Poison pills? Pellets of infection? Nothing so ordinary. The things are alive, they creep along the folds of the wing toward its base—they are, in short, young flies born at the instant the body of the mother fly struck the wing of the grasshopper. But a *young* fly would never be recognized as the offspring of its parent; it is a wormlike creature, or maggot, having neither wings nor legs and capable of moving only by extending and contracting its soft, flexible body (Fig. 182 D).

In form, the young *Sarcophaga kellyi* does not differ particularly from the maggots of other kinds of flies, but the Sarcophaga flies in general differ from most other insects in that their eggs are hatched within the bodies of the females, and these flies, therefore, give birth to young maggots instead of laying eggs. The female of *Sarcophaga kellyi*, then, when she launches her attack on the flying grasshopper, is munitioned with a load of young maggots ready to be discharged and stuck by the moisture of their bodies to the object of contact.   The young parasites thus palmed off by their mother on the grasshopper, who has no idea what has happened to him, make their way to the base of the wing of their unwitting host, where they find a tender membranous area which they penetrate and thereby enter the body of the victim.   Here they feed upon the liquids or tissues of the now helpless insect and grow to maturity in from ten to thirty days.   Meanwhile, however, the grasshopper has died, and when the parasites are full grown, they leave the dead body and bury themselves in the earth to a depth of from two to six inches. Here they undergo the transformation that will give them the form of their parents, and when they attain this stage they issue from the earth as adult winged flies.   Thus, one insect is destroyed that another may live.

Is the *Sarcophaga kellyi* a creature of uncanny shrewdness, an ingenious inventor of a novel way for avoiding the work of caring for her offspring?   Certainly her method is an improvement on that of leaving one's newborn progeny on a stranger's doorstep, for the victim of the fly must accept the responsibility thrust upon him whether he will or not.   But Doctor Kelly tells us that the flies do not know grasshoppers from other flying insects, such as moths and butterflies, in which their maggots do not find congenial hosts and never reach maturity.   Furthermore, he says, the ardent fly mothers will go after pieces of crumpled paper thrown into the wind and will discharge their maggots upon them, to which the helpless infants cling without hope of survival.   Such performances, and many similar ones that could be recounted of other insects, show that instinct is indeed blind and depends, not upon foresight, but on some mechanical action of the nervous system, which gives the desired result in the majority of cases but which is not guarded against unusual conditions or emergencies.

When we consider the many perfected instincts among insects, we are often shocked to find apparent cases of flagrant neglect on the part of nature for her creatures, where it would seem a remedy for their ills would be easy to supply.

In human society of modern times the criminal element has come to look no different from the law-abiding class of citizens. Formerly, if we may judge from pictures and stage representations, thieves and thugs were tough-looking individuals that could not be mistaken on sight, but

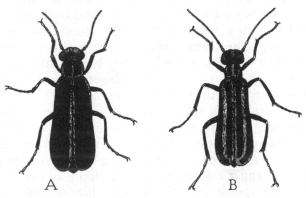

Fig. 11. Two blister beetles whose larvae feed on grasshopper eggs. (Twice natural size)
A, *Epicauta marginata*. B, *Epicauta vittata*

today our bandits are spruce young fellows that pass without suspicion in the crowd. And thus it is with the insects, all unsuspectingly one may be rubbing elbows with another that overnight will despoil his home, or that has already committed some act of violence against his neighbor. Here, for example, in the same field with the grasshoppers, is an innocent-looking beetle, about three-quarters of an inch in length, black and striped with yellow (Fig. 11 B). His entomological name is *Epicauta vittata*, which, of course, means nothing to a locust. He is now a vegetarian, but in his younger days he ravished the nest of a grasshopper and devoured the eggs, and his progeny will do the same again. Epicauta and others of his family

are known as "blister beetles" because they have a substance in their blood, called *cantharidin*, famous for its
blistering properties and formerly much used in medicine.
The female blister beetles of several species lay their eggs
in the ground in regions frequented by grasshoppers, where
the young on hatching can find the egg-pods of the latter.
The little beetles (Fig. 12) hatch in a form quite different
from that of their parents and are known as *triungulins*
because of two spines beside the single claw on each of
their feet, which gives the foot a three-clawed appearance.
Though the young scapegrace of a beetle is a housebreaker
and a thief, his story, like that of too many criminals,
unfortunately, makes interesting read-
ing, and the following account is taken,
with a few omissions, from the history
of *Epicauta vittata* as given by Dr.
C. V. Riley:

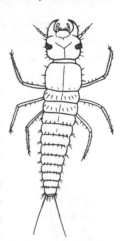

From July till the middle of October the
eggs are being laid in the ground in loose, irregular masses of about 130 on an average—the
female excavating a hole for the purpose, and
afterwards covering up the mass by scratching
with her feet. She lays at several different
intervals, producing in the aggregate probably
from four to five hundred ova. She prefers for
purposes of oviposition the very same warm
sunny locations chosen by the locusts, and
doubtless instinctively places her eggs near
those of these last, as I have on several occasions found them in close proximity. In the
course of about 10 days—more or less, according to the temperature of the ground—the
first larva or triungulin hatches. These little
triungulins (Fig. 12), at first feeble and per-

Fig. 12. The first-
stage larva, or "triun-
gulin," of the striped
blister beetle (fig. 11
B). Enlarged 12 times.
(From Riley)

fectly white, soon assume their natural light-brown color and commence
to move about. At night, or during cold or wet weather, all those
of a batch huddle together with little motion, but when warmed by
the sun they become very active, running with their long legs over the
ground, and prying with their large heads and strong jaws into every
crease and crevice in the soil, into which, in due time, they burrow

and hide. As becomes a carnivorous creature whose prey must be industriously sought, they display great powers of endurance, and will survive for a fortnight without food in a moderate temperature. Yet in the search for locust eggs many are, without doubt, doomed to perish, and only the more fortunate succeed in finding appropriate diet.

Reaching a locust egg-pod, our triungulin, by chance, or instinct, or both combined, commences to burrow through the mucous neck, or covering, and makes its first repast thereon. If it has been long in search, and its jaws are well hardened, it makes quick work through this porous and cellular matter, and at once gnaws away at an egg, first devouring a portion of the shell, and then, in the course of two or three days, sucking up the contents. Should two or more triungulins enter the same egg-pod, a deadly conflict sooner or later ensues until one alone remains the victorious possessor.

The surviving triungulin then attacks a second egg and more or less completely exhausts its contents, when, after about eight days from the time of its hatching, it ceases

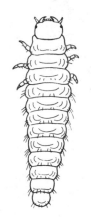

from its feeding and enters a period of rest. Soon the skin splits along the back, and the creature issues in the second stage of its existence. Very curiously, it is now quite different in appearance, being white and soft-bodied and having much shorter legs than before (Fig. 13). After feeding again on the eggs for about a week, the creature molts a second time and appears in a still different form. Then once more, and yet a fourth time, it sheds its skin and changes its form. Just before the fourth molt, however, it quits the eggs and burrows a short distance into the soil, where it composes itself for a period of retirement, and here undergoes

Fig. 13. The second-stage larva of the striped blister beetle. (From Riley)

another molt, in which the skin is not cast off. Thus the half-grown insect passes the winter, and in spring molts a sixth time and becomes active again, but not for long—its larval life is now about to close, and with another molt

it changes to a *pupa*, the stage in which it is to be transformed back into the form of its beetle parents. The final
change is accomplished in less than a week, and the
creature then emerges from the soil, now a fully-formed
striped blister beetle.

The grasshoppers' eggs furnish food for many other
insects besides the young blister beetles. There are species
of flies and of small wasplike insects whose larvae feed in
the egg-pods in much the same manner as do the triungulins, and there are still other species of general feeders
that devour the locust eggs as a part of their miscellaneous
diet. Notwithstanding all this destruction of the germs
of their future progeny, however, the grasshoppers still
thrive in abundance, for grasshoppers, like most other
insects, put their trust in the admonition that there is
safety in numbers. So many eggs are produced and stored
away in the ground each season that the whole force of
their enemies combined can not destroy them all, and
enough are sure to come through intact to render certain
the continuance of the species. Thus we see that nature
has various ways of accomplishing her ends—she might
have given the grasshopper eggs better protection in the
pods, but, being usually careless of individuals, she chose
to guarantee perpetuance with fertility.

CHAPTER II

THE GRASSHOPPER'S COUSINS

NATURE'S tendency is to produce groups rather than individuals. Any animal you can think of resembles in some way another animal or a number of other animals. An insect resembles on the one hand a shrimp or a crab, and on the other a centipede or a spider. Resemblances among animals are either superficial or fundamental. For example, a whale or a porpoise resembles a fish and lives the life of a fish, but has the skeleton and other organs of land-inhabiting mammals. Therefore, notwithstanding their form and aquatic habits, whales and porpoises are classed as mammals and not as fishes.

When resemblances between animals are of a fundamental nature, we believe that they represent actual blood relationships carried down from some far-distant common ancestor; but the determination of relationships between animals is not always an easy matter, because it is often difficult to know what are fundamental characters and what are superficial ones. It is a part of the work of zoologists, however, to investigate closely the structure of all animals and to establish their true relationships. The ideas of relationship which the zoologist deduces from his studies of the structure of animals are expressed in his classification of them. The primary divisions of the Animal Kingdom, which is generally likened to a tree, are called branches, or *phyla* (singular, *phylum*).

The insects, the centipedes, the spiders, and the shrimps, crayfish, lobsters, crabs, and other such creatures belong to the phylum *Arthropoda*. The name of this phylum means "jointed-legs"; but, since many other animals have jointed legs, the name is not distinctive, except in that the legs of the arthropods are particularly jointed, each being composed of a series of pieces that bend upon each other in different directions. A name, however, as everybody knows, does not have to mean anything, for Mr. Smith

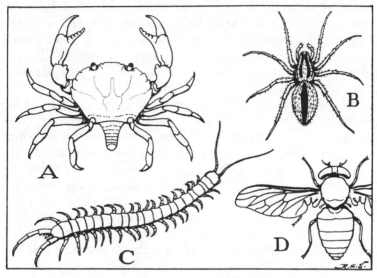

FIG. 14. Examples of four common classes of the Arthropoda
A, a crab (Crustacea). B, a spider (Arachnida). C, a centipede (Chilopoda).
D, a fly (Insecta, or Hexapoda)

may be a carpenter, and Mr. Carpenter a smith. A phylum is divided into *classes*, a class into *orders*, an order into *families*, a family into *genera* (singular, *genus*), and a genus is composed of *species* (the singular of which is also *species*). Species are hard to define, but they are what we ordinarily regard as the individual kinds of animals. Species are given double names, first the genus name, and second a specific name. For example, species of a common grasshopper genus named *Melanoplus* are distinguished as *Melanoplus atlanus*, *Melanoplus femur-rubrum*, *Melanoplus differentialis*, etc.

[27]

The insects belong to the class of the Arthropoda known as the *Insecta*, or *Hexapoda*. The word "insect," as we have seen, means "in-cut," while "hexapod" means "six-legged"—either term, then, doing very well for insects. The centipedes (Fig. 14 C) are the *Myriapoda*, or many-footed arthropods; the crabs (A), shrimps, lobsters, and others of their kind are the *Crustacea*, so called because most of them have hard shells; the spiders (B) are the *Arachnida*, named after that ancient Greek maiden so boastful of her spinning that Minerva turned her into a spider; but some arachnids, such as the scorpion, do not make webs.

The principal groups of insects are the orders. The grasshopper and its relatives constitute an order; the beetles are an order; the moths and butterflies are another order; the flies another; the wasps, bees, and ants still another. The grasshopper's order is called the *Orthoptera*, the word meaning "straight-wings," but, again, not significant in all cases, though serving very well as a name. The order is a group of related families, and, in the Orthoptera, the grasshoppers, or locusts, make one family, the katydids another, the crickets a third; and all these insects, together with some others less familiar, may be said to be the grasshopper's cousins.

The orthopteran families are notable in many ways, some for the great size attained by their members, some for their remarkable forms, and some for musical talent. While this chapter will be devoted principally to the cousins of the grasshopper, a few things of interest may still be said about the grasshopper himself, in addition to what was given in the preceding chapter.

THE GRASSHOPPER FAMILY

The family of the grasshoppers, or locusts, is the Acrididae. All the members are much alike in form and habits, though some have long wings and some short wings, and some reach the enormous size of nearly six inches in

PLATE 1

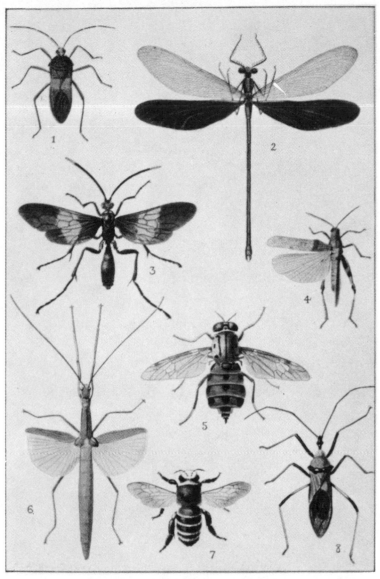

A group of insects representing five common entomological Orders. Figure 2 is a damselfly, a kind of dragonfly, from New Guinea, Order Odonata; 4 is a grasshopper, and 6 a winged walking-stick of Japan, representing two families of Orthoptera; 1 and 8 are sucking bugs, Order Hemiptera, which includes also the aphids and the cicadas; 3 is a wasp from Paraguay, and 7 a solitary bee from Chile, Order Hymenoptera; 5 is a two-winged fly of the Order Diptera, from Japan.

To entomologists these insects are known as follows: 1, *Paryphes laetus*; 2, unidentified; 3, *Pepsis completa*; 4, *Heliastus benjamini*; 5, *Pantophthalmus vittatus*; 6, *Micadina phluctanoides*; 7, *Caupolicana fulvicollis*; 8, *Margasus afzeli*

length. The front wings are long and narrow (Fig. 63, W_2), somewhat stiff, and of a leathery texture. They are laid over the thinner hind wings as a protection to the latter when the wings are folded over the back, and for this reason they are called the *tegmina* (singular, *tegmen*). The hind wings, when spread (W_3), are seen to be large fans, each with many ribs, or veins, springing from the base. These wings are gliders rather than organs of flight. For most grasshoppers leap into the air by means of their strong hind legs and then sail off on the outspread wings as far as a weak fluttering of the latter will carry them. One of our common species, however, the Carolina locust (Frontispiece), is a strong flyer, and when

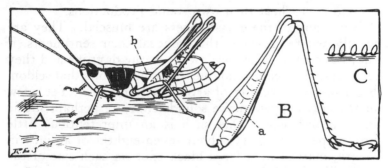

Fig. 15. A grasshopper, *Chloealtis conspersa*, that makes a sound by scraping its hind thighs over sharp-edged veins of its wings

A, the male grasshopper, showing the sound-making veins of the wing (*b*). B, inner surface of right hind leg, showing row of teeth (*a*) on the femur. C, several teeth of the femur (enlarged)

flushed flits away on an undulating course over the weeds and bushes and sometimes over the tops of small trees, but always swerving this way and that as if undecided where to alight. The great flights of the migratory locusts, described in the last chapter, are said to have been accomplished more by the winds than by the insects' strength of wing.

The locusts are distinguished by the possession of large

organs on the sides of the body that appear to be designed for purposes of hearing. No insect, of course, has "ears" on its head; the grasshopper's supposed hearing organs are located on the base of the abdomen, one on each side (Fig. 63, *Tm*). Each consists of an oval depression of the body wall with a thin eardrumlike membrane, or *tympanum*, stretched over it. Air sacs lie against the inner face of the membrane, furnishing the equilibrium of air pressure necessary for free vibration in response to sound waves, and a complicated sensory apparatus is attached to its inner wall. Even with such large ears, however, attempts at making the grasshopper hear are never very successful; but its tympanal organs have the same structure as those of insects noted for their singing, which presumably, therefore, can hear their own sound productions.

Not many of the grasshoppers are muscial. They are mostly sedate creatures that conceal their sentiments, if they have any. They are awake in the daytime and they sleep at night—commendable traits, but habits that seldom beget much in the way of artistic attainment. Yet a few of the grasshoppers make sounds that are perhaps music in their own ears. One such is an unpretentious little brown species (Fig. 15) about seven-eighths of an inch in length, marked by a large black spot on each side of the saddlelike shield that covers his back between the head and the wings. He has no other name than his scientific one of *Chloealtis conspersa*, for he is not widely known, since his music is of a very feeble sort. According to Scudder, his only notes resemble *tsikk-tsikk-tsikk*, repeated ten or twelve times in about three seconds in the sun, but at a slightly lower rate in the shade. Chloealtis is a fiddler and plays two instruments at once. The fiddles are his front wings, and the bows his hind legs. On the inner surface of each hind thigh, or *femur*, there is a row of minute teeth (Fig. 15 B, *a*), shown more magnified at C. When the thighs are rubbed over the edges of the wings, their teeth scrape on a sharp-edged vein indicated by *b*. This produces the *tsikk*-sound just mentioned. Such notes contain little music to us, but Scudder says he has seen three males singing to one female at the same time. This female, however,

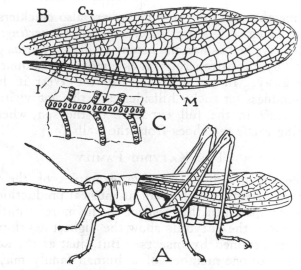

FIG. 16. A grasshopper, *Mecostethus gracilis*, that makes a sound by scraping sharp ridges on the inner surfaces of its hind thighs over toothed veins of the wings

A, the male grasshopper. B, left front wing; the rasping vein is the one marked *I*. C, a part of the rasping vein and its branches more enlarged, showing rows of teeth

was busy laying her eggs in a near-by stump, and there is no evidence given to show that even she appreciated the efforts of her serenaders.

Several other little grasshoppers fiddle after the manner of Chloealtis; but another, *Mecostethus gracilis* by name (Fig. 16), instead of having the rasping points on the legs, has on each fore wing one vein (B, *I*) and its branches provided with many small teeth, shown enlarged at C, upon which it scrapes a sharp ridge situated on the inner surface of the hind thigh.

In another group of grasshoppers there are certain species that make a noise as they fly, a crackling sound apparently produced in some way by the wings themselves. One of these, common through the Northern States, is known as the cracker locust, *Circotettix verruculatus*, on account of the loud snapping notes it emits. Several other members of the same genus are also cracklers, the noisiest being a western species called *C. carlingianus*. Scudder says he has had his attention drawn to this grasshopper "by its obstreperous crackle more than a quarter of a mile away. In the arid parts of the West it has a great fondness for rocky hillsides and the hot vicinity of abrupt cliffs in the full exposure to the sun, where its clattering rattle re-echoes from the walls."

The Katydid Family

While the grasshoppers give examples of the more primitive attempts of insects at musical production and may be compared in this respect to the more primitive of human races, the katydids show the highest development of the art attained by insects. But, just as the accomplishments of one member of a human family may give prestige to all his relations and descendants, so the talent of one noted member of the katydid family has given notoriety to all his congeners, and his justly deserved name has come to be applied by the undiscriminating public to a whole tribe of singers of lesser or very mediocre talent whose only claim to the name of katydid is that of family relationship. In Europe the katydids are called simply the longhorn grasshoppers. In entomology the family is now the Tettigoniidae, though it had long been known as the Locustidae.

The katydids in general are most easily distinguished from the locusts, or shorthorn grasshoppers, by the great length of their antennae, those delicate, sensitive, tapering threads projecting from the forehead. But the two families differ also in the number of joints in their feet, the grasshoppers having three (Fig. 17 A) and the katydids four (B). The grasshoppers place the entire foot on the ground, while the katydids ordinarily walk on the three basal segments only, carrying the long terminal joint elevated. The basal segments have pads on their under sides that adhere to any smooth surface such as that of a leaf, but the terminal joint bears a pair of claws used when it is necessary to grasp the edge of a support. The katydids are mostly creatures of the night and, though usually plain green in color, many of them have elegant forms. Their attitudes and general comportment suggest much more refinement and a higher breeding than that of the heavy-bodied locusts. Though some members of the katydid family live in the fields and are very grasshopperlike or even cricketlike in form and manners, the characteristic species are seclusive inhabitants of shrubbery or trees. These are the true aristocrats of the Orthoptera.

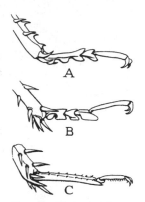

FIG. 17. Distinctive characters in the feet of the three families of singing Orthoptera

A, hind foot of a grasshopper. B, hind foot of a katydid. C, hind foot of a cricket

An insect musician differs in many respects from a human musician, aside from that of being an insect instead of a human being. The insect artists are all instrumentalists; but since the poets and other ignorant people always speak of the "singing" of the crickets and katydids, it will be easier to use the language of the public than to correct it, especially since we have nothing better to offer than the word *stridulating*, a Latin derivative meaning "to creak." But words do not matter if we explain what we mean by them. It must be understood, therefore, that though we speak of the "songs" of insects, insects do not have true voices in the sense that "voice" is the production of sound by the breath playing on vocal cords. All the musical instruments of insects, it is true, are parts of their bodies; but they are to be likened to fiddles or drums, since, for the production of sound, they depend upon rasping and vibrating surfaces. The rasping surfaces are usually, as in the instruments of the grasshoppers (Figs. 15, 16), parts of the legs and the wings. The sound may be intensified, as in the body of a stringed instrument, by special resonating areas, sometimes on the wings, sometimes on the body. The cicadas, a group of musical insects to be described in a special chapter, have large drumheads in the wall of the body with which they produce their shrill music. They do not beat these drums, but cause them to vibrate by muscles in the body. The musical members of the insect families are in nearly all cases the males, and it is usually supposed that they give their concerts for the purpose of engaging the females, but that this is so in all cases we can not be certain.

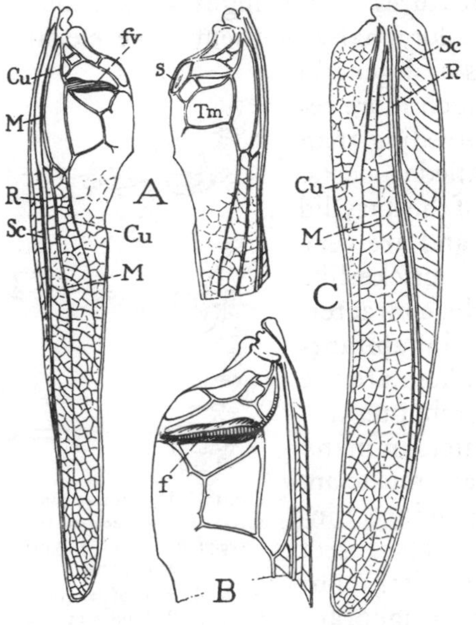

The musical instruments of the katydids are quite different from those of the grasshoppers, being situated on the over-

FIG. 18. The front wings, or tegmina, of a meadow grasshopper, *Orchelimum laticauda*, illustrating the sound-making organs typical of the katydid family

A, left front wing and basal part of right wing of male, showing the four main veins: subcosta (*Sc*), radius (*R*), media (*M*), and cubitus (*Cu*); also the enlarged basal vibrating area, or tympanum (*Tm*), of each wing, the thick file vein (*fv*) on the left, and the scraper (*s*) on the right

B, lower surface of base of left wing of male, showing the file (*f*) on under side of the file vein (A, *fv*)

C, right front wing of female, which has no sound-making organs, showing simple normal venation

[34]

lapping bases of the front wings, or tegmina. On this account the front wings of the males are always different from those of the females, the latter retaining the usual or primitive structure. The right wing of a female in one of the more grasshopperlike species, *Orchelimum laticauda* (Fig. 30), is shown at C of Figure 18. The wing is traversed by four principal veins springing from the base. The one nearest the inner edge is called the *cubitus* (*Cu*) and the space between it and this margin of the wing is filled with a network of small veins having no particular arrangement. In the wings of the male, however, shown at A of the same figure, this inner basal field is much enlarged and consists of a thin, crisp membrane (*Tm*), braced by a number of veins branching from the cubitus (*Cu*). One of these (*fv*), running crosswise through the membrane, is very thick on the left wing, and when the wing is turned over (B) it is seen to have a close series of small crossridges on its under surface which convert it into

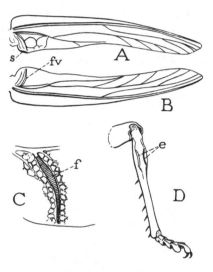

Fig. 19. Wings, sound-making organs, and the "ears" of a conehead grasshopper, *Neoconocephalus ensiger*, a member of the katydid family

A, B, right and left wings, showing the scraper (*s*) on the right, and the file vein (*fv*) on the left. C, under surface of the file vein, showing the file (*f*). D, front leg, showing slits (*e*) on the tibia opening into pockets containing the hearing organs (fig. 20 A)

a veritable file (*f*). On the right wing this same vein is much more slender and its file is very weak, but on the basal angle of this wing there is a stiff ridge (*s*) not developed on the other. The katydids always fold the wings with the *left overlapping the right*, and in this position the file of the former lies above the ridge (*s*) of the latter. If now the wings are moved sidewise, the *file* grating on the ridge or *scraper* causes a rasping sound, and this is the way the katydid makes the notes of its music. The tone and volume of the sound, however, are probably in large part produced by the vibration of the thin basal membranes of the wings, which are called the *tympana* (*Tm*).

The instruments of different players differ somewhat in the details of their structure. There are variations in the form and size of the file and the scraper on the wings of different species, and differences in the veins supporting the tympanal areas, as shown in the drawings of these parts from a conehead (Fig. 27) given at A, B, and C, of Figure 19. In the true katydid, the greatest singer of the family, the file, the scraper, the tympana, and the wings themselves (Fig. 26) are all very highly developed to form an instrument of great efficiency. But, in general, the instruments of different species do not differ nearly so much as do the notes produced from them by their owners. An endless number of tunes may be played upon the same fiddle. With the insects each musician knows only one tune, or a few simple variations of it, and this he has inherited from his ancestors along with a knowledge of how to play it on his inherited instrument. The stridulating organs are not functionally developed until maturity, and then the insect forthwith plays his native air. He never disturbs the neighbors with doleful notes while learning.

Very curiously, none of the katydids nor any member of their family has the earlike organs on the sides of the body possessed by the locusts. What are commonly supposed to be their organs of hearing are located in their front legs, as are the similar organs of the crickets. Two vertical slits on the upper parts of the shins, or *tibiae* (Fig. 19 D, *e*), open each into a small pocket (Fig. 20 A, *E*) with a tympanumlike membrane (*Tm*) stretched across its inner wall. Between the membranes are air cavities (*Tra*) and a complicated sensory receptive apparatus (B) connected by a nerve through the basal part of the leg with the central nervous system.

There are several groups of katydids, classed as sub-

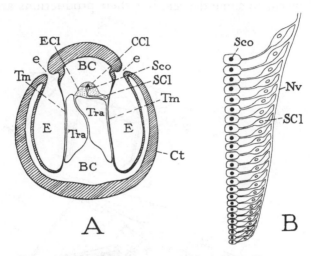

Fig. 20.   The probable auditory organ of the front leg of *Decticus*, a member of the katydid family.   (Simplified from Schwabe)

A, cross-section of the leg through the auditory organ, showing the ear slits (*e, e*) leading into the large ear cavities (*E, E*) with the tympana (*Tm, Tm*) on their inner faces.   Between the tympana are two tracheae (*Tra, Tra*) dividing the leg cavity into an upper and a lower channel (*BC, BC*).   The sensory apparatus forms a crest on the outer surface of the inner trachea, each element consisting of a cap cell (*CCl*), an enveloping cell (*ECl*) containing a sense rod (*Sco*), and a sense cell (*SCl*).   *Ct*, the thick cuticula forming the hard wall of the leg

B, surface view of the sensory organ, showing the elements graded in size from above downward.   The sense cells (*SCl*) are attached to the nerve (*Nv*) along the inner side of the leg

families.   A subfamily name ends in *inae* to distinguish it from a family name, which, after the Latin fashion, terminates in *idae*.

THE ROUND-HEADED KATYDIDS

The members of this first group of the katydid family are characterized by having large wings and a smooth round forehead. They compose the subfamily Phaneropterinae, which includes species that attain the acme of grace, elegance, and refinement to be found in the entire orthopteran order. Nearly all the round-headed katydids are musical to some degree, but their productions are not

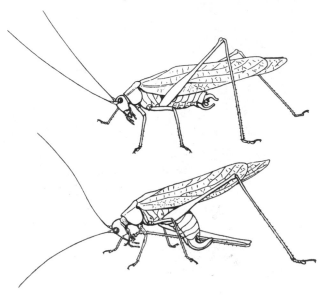

Fig. 21.   A bush katydid, *Scudderia furcata*
Upper figure, a male; lower, a female in the act of cleaning a
hind foot

of a high order. On the other hand, though their notes are in a high key, they are usually not loud and not of the kind that keep you awake at night.

Among this group are the bush katydids, the species of which are of medium size with slenderer wings than the others, and are comprised in the genus usually known as *Scudderia* but also called *Phaneroptera*. They have acquired the name of bush katydids because they are usually found on low shrubbery, particularly along the edges of moist meadows, though they inhabit other places, too, and their notes are often heard at night about the house. Our commonest species, and one that occurs over most of the United States, is the fork-tailed bush katydid (*Scudderia furcata*). Figure 21 shows a male and a female, the female in the act of cleaning the pads on one of her hind feet. The katydids are all very particular about keeping their feet clean, for it is quite essential to have their adhesive pads always in perfect working order; but they are so continually stopping whatever they may be doing to lick one foot or another, like a dog scratching fleas, that it looks more like an ingrown habit with them than a necessary act of cleanliness. The fork-tailed katydid is an unpretentious singer and has only one note, a high-pitched *zeep* reiterated several times in succession. But it does not repeat the series continuously, as most other singers do, and its music is likely to be lost to human ears in the general din from the jazzing bands of crickets. Yet occasionally its soft *zeep, zeep, zeep* may be heard from a near-by bush or from the lower branches of a tree.

The notes of other species have been described as *zikk, zikk, zikk*, or *zeet, zeet, zeet*, and some observers have recorded two notes for the same species. Thus Scudder says that the day notes and the night notes of *Scudderia curvicauda* differ considerably, the day note being represented by *bzrwi*, the night note, which is only half as long as the other, by *tchw*. (With a little practice the reader should be able to give a good imitation of this katydid.) Scudder furthermore says that they change from the day note to the night note when a cloud passes over the sun as they are singing by day.

The genus *Amblycorypha* includes a group of species having wider wings than those of the bush katydids. Most of them are indifferent singers; but one, the oblong-winged katydid (*A. oblongifolia*), found over all the eastern half of the United States and southern Canada, is noted for its large size and dignified manners. A male (Fig. 22), kept by the writer one summer in a cage, never once lost his decorum by the humiliation of confinement. He lived apparently a natural and contented life, feeding on grape leaves and on ripe grapes, obtaining the pulp of the latter by gnawing holes through the skin. He was always sedate, always composed, his motions always slow and deliberate. In walking he carefully lifted each foot and brought the leg forward with a steady movement to the new position,

FIG. 22. The oblong-winged katydid, *Amblycorypha oblongifolia*, male

where the foot was carefully set down again. Only in the act of jumping did he ever make a quick movement of any sort. But his preparations for the leap were as calm and unhurried as his other acts: pointing the head upward, dipping the abdomen slowly downward, the two long hind legs bending up in a sharp inverted V on each side of the body, he would lead one to think he was deliberately preparing to sit down on a tack; but, all at once, a catch seems to be released somewhere as he suddenly springs upward into the leaves overhead at which he had taken such long and careful aim.

For a long time the aristocratic prisoner uttered no sound, but at last one evening he repeated three times a squeaking note resembling *shriek* with the *s* much aspirated and with a prolonged vibration on the *ie*. The next evening he played again, making at first a weak *swish*, *swish*, *swish*, with the *s* very sibilant and the *i* very vibratory. But after giving this as a prelude he began a shrill *shrie-e-e-e-k*, *shrie-e-e-e-k*, repeated six times, a loud sound described by Blatchley as a "creaking squawk—like the noise made by drawing a fine-toothed comb over a taut string."

The best-known members of the round-headed katydids, and perhaps of the whole family, are the angular-winged katydids (Fig. 23). These are large, maple-leaf green insects, much flattened from side to side, with the leaflike wings folded high over the back and abruptly bent on their upper margins, giving the creatures the humpbacked appearance from which they get their name of angular-winged katydids. The sloping surface of the back in front of the hump makes a large flat triangle, plain in the female, but in the male corrugated and roughened by the veins of the musical apparatus.

There are two species of the angular-winged katydids in the United States, both belonging to the genus *Microcentrum*, one distinguished as the larger angular-winged katydid, *M. rhombifolium*, and the other as the smaller angular-winged katydid, *M. retinerve*. The females of the larger species (Fig. 23), which is the more common one, reach a length of $2\frac{3}{8}$ inches measured to the tips of the wings. They lay flat, oval eggs, stuck in rows overlapping like scales along the surface of some twig or on the edge of a leaf.

The angular-winged katydids are attracted to lights and may frequently be found on warm summer nights in the shrubbery about the house, or even on the porch and the screen doors. Members of the larger species usually make their presence known by their soft but high-pitched notes resembling *tzeet* uttered in short series, the first notes repeated rapidly, the others successively more slowly as the tone becomes also less sharp and piercing. The song may be written *tzeet-tzeet-tzeet-tzeet-tzek-tzek-tzek-tzuk-tzuk*, though the high key and shrill tones of the notes must be

FIG. 23. The larger angular-winged katydid, *Microcentrum rhombifolium*
Upper figure, a male; lower, a female

imagined. Riley describes the song as a series of raspings "as of a stiff quill drawn across a coarse file," and Allard says the notes "are sharp, snapping crepitations and sound like the slow snapping of the teeth of a stiff comb as some object is slowly drawn across it." He represents them thus: *tek-ek-ek-ek-ek-ek-ek-ek-ek-ek-tzip*. But, however the song of Microcentrum is to be translated into English, it contains no suggestion of the notes of his famous cousin, the true katydid. Yet most people confuse the two species, or rather, hearing the one and seeing the other, they draw the obvious but erroneous conclusion that the one seen makes the sounds that are heard.

The smaller angular-winged katydid, *Microcentrum retinerve*, is not so frequently seen as the other, but it has similar habits, and may be heard in the vines or shrubbery about the house at night. Its song is a sharp *zeet, zeet, zeet*, the three syllables spaced as in *ka-ty-did*, and it is probable that many people mistake these notes for those of the true katydid.

The angular-winged katydids are very gentle and unsuspicious creatures, allowing themselves to be picked up without any attempt at escaping. But they are good flyers, and when launched into the air sail about like miniature airplanes, with their large wings spread out straight on each side. When at rest they have a comical habit of leaning over sidewise as if their flat forms were top-heavy.

THE TRUE KATYDID

We now come to that artist who bears by right the name of "katydid," the insect (Fig. 24) known to science as *Pterophylla camellifolia* and to the American public as the greatest of insect singers. Whether the katydid is really a musician or not, of course, depends upon the critic, but of his fame there can be no question, for his name is a household term as familiar as that of any of our own great artists, notwithstanding that there is no phonographic record of his music. To be sure, the cicada has more of a world-wide reputation than the katydid, for he has representatives in many lands, but he has not put his song into words the public can understand. And if simplicity be the test of true art, the song of the katydid stands the test, for nothing could be simpler than merely *katy-did*, or its easy variations, such as *katy*, *katy-she-did*, and *katy-didn't*.

Yet though the music of the katydid is known by ear or by reputation to almost every native American, few of us

FIG. 24. The true katydid, *Pterophylla camellifolia*, a male

are acquainted with the musician himself. This is because he almost invariably chooses the tops of the tallest trees for his stage and seldom descends from it. His lofty platform, moreover, is also his studio, his home, and his world, and the reporter who would have a personal interview must be efficient in tree climbing. Occasionally, though, it happens that a singer may be located in a smaller tree where access to him is easier or from which he may be dislodged by shaking. A specimen, secured in this way on August 12, lived till October 18 and furnished material for the following notes:

The physical characters of the captive and some of his attitudes are shown in Figures 24 and 25. His length is 1¾ inches from the forehead to the tips of the folded wings; the front legs are longer and thicker than in most other members of the family, while the hind legs are unusually short. The antennae, though, are extremely long, slender, and very delicate filaments, $2^{13}/_{16}$ inches in length.

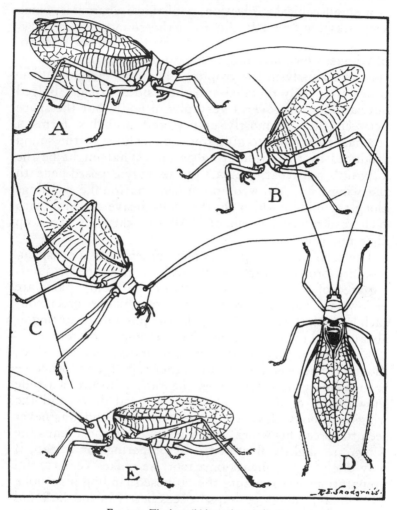

FIG. 25.   The katydid in various attitudes
A, usual position of a male while singing.   B, attitude while running rapidly on
a smooth surface.   C, preparing to leap from a vertical surface.   D, a male,
seen from above, showing the stridulating area at the base of the wings.   E, a
female, showing the broad, flat, curved ovipositor

Between the bases of the antennae on the forehead there is a small conical projection, a physical character which separates the true katydid from the round-headed katydids and assigns him to the subfamily called the Pseudophyllinae, which includes, besides our species, many others that live mostly in the tropics. The rear margins of the wings are evenly rounded and their sides strongly bulged outward as if to cover a very plump body, but the space between them is mostly empty and probably forms a resonance chamber to give tone and volume to the sound produced by the stridulating parts. What might be the katydid's waistcoat, the part of the body exposed beneath the wings, has a row of prominent buttonlike swellings along the middle which rhythmically heave and sink with each respiratory movement. All the katydids are deep abdominal breathers.

The color of the katydid is plain green, with a conspicuous dark-brown triangle on the back covering the stridulating area of the wings. The tips of the mouth parts are yellowish. The eyes are of a pale transparent green, but each has a dark center which, like the pupil in a painting, is always fixed upon you from whatever angle you retreat.

The movements of the captive individual are slow, though in the open he can run rather rapidly, and when he is in a hurry he often takes the rather absurd attitude shown at B of Figure 25, with the head down and the wings and body elevated. He never flies, and was never seen to spread his wings, but when making short leaps the wings are slightly fluttered. In preparing for a leap, if only one of a few inches or a foot, he makes very careful preparations, scrutinizing the proposed landing place long and closely, though perhaps he sees better in the dark and acts then with more agility. If the leap is to be made from a horizontal surface, he slowly crouches with the legs drawn together, assuming an attitude more familiar in a cat; but, if the jump is to be from a vertical support, he raises himself on his long front legs as at C of Figure 25, suggesting a camel browsing on the leaves of a tree. He sparingly eats leaves of oak and maple supplied to him in his cage, but appears to prefer fresh fruit and grapes, and relishes bread soaked in water. He drinks rather less than most orthopterons.

When the katydids are singing at night in the woods they appear to be most wary of disturbance, and often the voice of a person approaching or a crackle underfoot is sufficient to quiet a singer far overhead. The male in the cage never utters a note until he has been in darkness and quiet for a considerable time. But when he seems to be assured of solitude he starts his music, a sound of tremendous volume in a room, the tones incredibly harsh and rasping at close range, lacking entirely that melody they acquire with space and distance. It is only by extreme caution that the performer may be approached while singing, and even then the brief flash of a light is usually enough to silence those stentorian notes. Yet occasionally a glimpse may be had of the musician as he plays, most frequently standing head downward, the body braced rather stiffly on the legs, the front wings only slightly elevated, the tips of the hind wings projecting a little from between them, the abdomen depressed and breathing strongly, the long antennal threads waving about in all directions. Each syllable appears to be produced by a separate series of vibrations made by a rapid shuffling of the wings, the middle one being more hurried and the last more conclusively stressed, thus producing the sound so suggestive of *ka-ty-did'*, *ka-ty-did'*, which is repeated regularly about sixty times a minute on warm nights. Usually at the start, and often for some time, only two notes are uttered, *ka-ty*, as if the player has difficulty in falling at once into the full swing of *ka-ty-did*.

The structure of the wings and the details of the stridulating parts are shown in Figure 26. The wings (A, B) fold vertically against the sides of the body, but their inner basal parts form wide, stiff, horizontal, triangular flaps that overlap, the left on top of the right. A thick, sunken, crosswise vein (*fv*) at the base of the left tympanum (*Tm*) is the file vein. It is shown from below at C where the broad, heavy file (*f*) is seen with its row of extremely coarse rasping ridges. The same vein on the right wing (**B**) is much smaller and has no file, but the inner basal angle of the tympanum is produced into a large lobe bearing a strong scraper (*s*) on its margin.

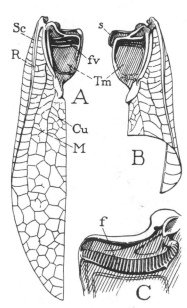

The quality of the katydid's song seems to differ somewhat in different parts of the country. In the vicinity of Washington, the insects certainly say *ka-ty-did* as plainly as any insect could. Of course, the sound is more literally to be represented as *kă ki-kăk'*, accented on the last syllable. When only two syllables are pronounced they are always the first two. Sometimes an individual in a band utters four syllables, "katy-she-did" or *kă ki-kă-kăk'*, and again a whole band is heard singing in four notes with only an occasional singer giving three. It is said that in certain parts of the South the katydid is called a "cackle-jack," a name which, it must be admitted, is a very literal translation of the notes, but one lacking in sentiment and unbefitting an artist of such repute. In New England, the katydids heard by the writer in Connecticut and in the western part of Massachusetts uttered only two syllables much

FIG. 26. Wings and the sound-making organs of the male katydid

A, left front wing, showing the greatly enlarged tympanal area (*Tm*), with its thick file vein (*fv*). B, base of right fore wing, with large scraper (*s*) on its inner angle, but with a very small file vein. C, under surface of file vein of left wing, showing the large, flat, coarsely-ribbed file (*f*)

more commonly than three, and the sounds were extremely harsh and rasping, being a loud *squă-wăk′*, *squă-wăk′*, *squă-wăk′*, the second syllable a little longer than the first. This is not the case with those that say *ka-ty*. When there were three syllables the series was *squă-wă-wăk′*. If all New England katydids sing thus, it is not surprising that some New England writers have failed to see how the insects ever got the name of "katydid." Scudder says "their notes have a shocking lack of melody"; he represents the sound by *xr*, and records that the song is usually of only two syllables. "That is," he says, " they rasp their fore wings twice rather than thrice; these two notes are of equal (and extraordinary) emphasis, the latter about one-quarter longer than the former; or if three notes are given, the first and second are alike and a little shorter than the last."

When we listen to insects singing, the question always arises of why they sing, and we might as well admit that we do not know what motive impels them. It is probably an instinct with males to use their stridulating organs, but in many cases the tones emitted are clearly modified by the physical or emotional state of the player. The music seems in some way to be connected with the mating of the sexes, and the usual idea is that the sounds are attractive to the females. With many of the crickets, however, the real attraction that the male has for the female is a liquid exuded on his back, the song apparently being a mere advertisement of his wares. In any case the ecstacies of love and passion ascribed to male insects in connection with their music are probably more fanciful than real. The subject is an enchanted field wherein the scientist has most often weakened and wandered from the narrow path of observed facts, and where he has indulged in a freedom of imagination permissible to a poet or to a newspaper reporter who wishes to enliven his chronicle of some event in the daily news, but which does not contribute anything substantial to our knowledge of the truth.

THE CONEHEADS

This group of the katydid family contains slender, grasshopperlike insects that have the forehead produced into a large cone and the face strongly receding, but which also possess long, slender antennae that distinguish them from the true or shorthorn grasshoppers. They constitute the subfamily Copiphorinae.

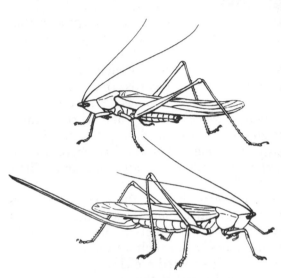

FIG. 27. A conehead grasshopper, or katydid, *Neoconocephalus retusus*
Upper figure, a male; lower, a female, with extremely long ovipositor

One of the commonest and most widely distributed of the larger coneheads is the species known as *Neoconocephalus ensiger*, or the "sword-bearing conehead." It is the female, however, that carries the sword; and it is not a sword either, but merely the immensely long egg-laying instrument properly called the *ovipositor*. The female conehead shown at B of Figure 27, has a similar organ, though she belongs to a species called *retusus*. The two species are very similar in all respects except for slight differences in the shape of the cone on the head. They look like slim, sharp-headed grasshoppers, 1½ to 1¾ inches in length, usually bright green in color, though sometimes brown.

The song of *ensiger* sounds like the noise of a miniature sewing machine, consisting merely of a long series of one note, *tick, tick, tick, tick,* etc., repeated indefinitely. Scudder says *ensiger* begins with a note like *brw*, then pauses an instant and immediately emits a rapid succession of sounds like *chwi* at the rate of about five per second and continues them an unlimited time. McNeil represents the notes as *zip, zip, zip;* Davis expresses them as *ik, ik, ik;* and Allard hears them as *tsip, tsip, tsip*. The song of *retusus* (Fig. 27) is quite different. It consists of a long shrill whir which Rehn and Hebard describe as a continuous *zeeeeeeeee*. The sound is not loud but is in a very high key and rises in pitch as the player gains speed in his wing movements, till to some human ears it becomes almost inaudible, though to others it is a plain and distinct screech.

A large conehead and one with a much stronger instrument is the robust conehead, *Neoconocephalus robustus* (Fig. 28). He is one of the loudest singers of North American Orthoptera, his song being an intense, continuous buzz, somewhat resembling that of a cicada. A caged specimen singing in a room makes a deafening noise.

Fig. 28. The robust conehead, *Neoconocephalus robustus*, in position of singing, with fore wings separated and somewhat elevated, the head downward

The principal buzzing sound is accompanied by a lower, droning hum, the origin of which is not clear, but which is probably some secondary vibration of the wings. The player always sits head downward while performing, and the breathing motions of the abdomen are very deep and rapid. The robust conehead is an inhabitant of dry, sandy places along the Atlantic coast from Massachusetts to Virginia and, according to Blatchley, of similar places near the shores of Lake Michigan in Indiana. The writer made its acquaintance in Connecticut on the sandy flats of the Quinnipiac Valley, north of New Haven, where its shrill song may be heard on summer nights from long distances.

THE MEADOW GRASSHOPPERS

These are trim, slim little grasshopperlike insects, active by day, that live in moist meadows where the vegetation is always fresh and juicy. They constitute the subfamily Conocephalinae of the katydid family, having conical

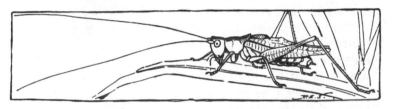

FIG. 29. The common meadow grasshopper, *Orchelimum vulgare*, a member of the katydid family

heads like the last group, but being mostly of smaller size. There are numerous species of the meadow grasshoppers, but most of them in the eastern part of the United States belong to two genera known as *Orchelimum* and *Conocephalus*. The most abundant and most widely distributed member of the first is the common meadow grasshopper, *Orchelimum vulgare*. A male is shown in Figure 29. He is a little over an inch in length, with head rather large for his size and with big eyes of a bright orange color. The ground color of his body is greenish, but the top of the head and the thoracic shield is occupied by a long triangular dark-brown patch, while the stridulating area of

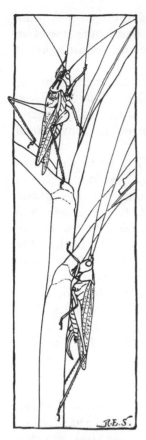

the wings is marked by a brown spot at each corner. These little grasshoppers readily sing in confinement, both in the day and at night. Their music is very unpretentious and might easily be lost out of doors, consisting mostly of a soft, rustling buzz that lasts two or three seconds. Often the buzz is preceded or followed by a series of clicks made by a slower movement of the wings. Frequently the player opens the wings for the start of the song with a single click, then proceeds with the buzz, and finally closes with a few slow movements that produce the concluding series of clicks. But very commonly he gives only the buzz without prelude or staccato ending.

Another common member of the genus is the agile meadow grasshopper, *Orchelimum agile*. Its music is said to be a long *zip*, *zip*, *zip*, *zee-e-e-e*, with the *zip* syllable repeated many times. These two elements, the *zip* and *zee*, are characteristic of the songs of all the Orchelimums, some giving more stress to the first and others to the second, and

FIG. 30. The handsome meadow grasshopper, *Orchelimum laticauda*

Upper figure, a male; lower, a female

FIG. 31. The slender meadow grasshopper, *Conocephalus fasciatus*, one of the smallest members of the katydid family

[53]

sometimes either one or the other is omitted. A very pretty species of the genus is the handsome meadow grasshopper, *Orchelimum laticauda* (or *pulchellum*) shown in Figure 30. When at rest, both males and females usually sit close to a stem or leaf with the middle of the body in contact with the support and the long hind legs stretched out behind. Davis says the song of this species is a *zip, zip, zip, z, z, z,* quite distinguishable from that of *O. vulgare.*

Still smaller meadow grasshoppers belong to the genus *Conocephalus,* more commonly called *Xiphidium.* One of the most abundant species, the slender meadow grasshopper, *C. fasciatus,* is shown in Figure 31. It is less than an inch in length, the body green, the back of the thorax dark brown, the wings reddish-brown, and the back of the abdomen marked with a broad brown stripe. Allard says the song of this little meadow grasshopper may be expressed as *tip, tip, tip, tseeeeeeeeeeeeee,* but that the entire song is so faint as almost to escape the hearing. Piers describes it as *ple-e-e-e-e-e, tzit, tzit, tzit, tzit.* Like the song of *Orchelimum vulgare* it apparently may either begin or end with staccato notes.

THE SHIELD BEARERS

Another large group of the katydid family is the subfamily Decticinae, mostly cricketlike insects that live on the ground, but which have wings so short (Fig. 32) that they are poor musicians. They are called "shield bearers" because the large back plate of the first body segment is more or less prolonged like a shield over the back. Most of the species live in the western parts of the United States, where the individuals sometimes become so abundant as to form large and very destructive bands. One such species is the Mormon cricket, *Anabrus simplex,* and another is the Coulee cricket, *Peranabrus scabricollis* (Fig. 32), of the dry central region of the State of Washington. The females of these species are commonly wingless, but the males have short stubs of front wings that retain the stridulating organs and enable them to sing with a brisk chirp.

Still another large subfamily of the Tettigoniidae is the

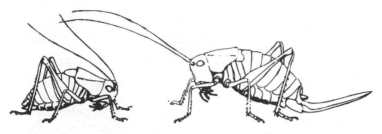

FIG. 32. The Coulee cricket, *Peranabrus scabricollis*, male and female, an example of a cricketlike member of the katydid family

Rhadophorinae, including the insects known as "camel crickets." But these are all wingless, and therefore silent.

THE CRICKET FAMILY

The chirp of the cricket is probably the most familiar note of all orthopteran music. But the only cricket commonly known to the public is the black field cricket, the lively chirper of our yards and gardens. His European cousin, the house cricket, is famous as the "cricket on the hearth" on account of his fondness for fireside warmth which so stimulates him that he must express his animation in song. This house cricket has been known as Gryllus since the time of the ancient Greeks and Romans, and his name has been made the basis for the name of his family, the Gryllidae, for there are numerous other crickets, some that live in trees, some in shrubbery, some on the ground, and others in the earth.

The crickets have long slender antennae like those of the katydids, and also stridulating organs on the bases of the wings, and ears in their front legs. But they differ from the katydids in having only three joints in their feet (Fig. 17 C). The cricket's foot in this respect resembles the foot of the grasshopper (A), but usually differs from that of the grasshopper in having the basal joint smooth or hairy all around or with only one pad on the under surface. In most crickets, also, the second joint of the foot is very small.

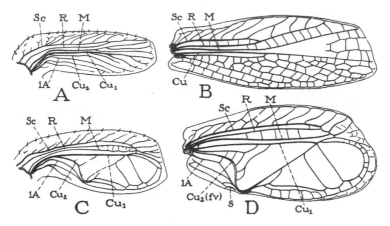

FIG. 33. The wings of a tree cricket

A, right front wing of an immature female, showing normal arrangement of veins: *Sc*, subcosta; *R*, radius; *M*, media; *Cu₁*, first branch of cubitus; *Cu₂*, second branch of cubitus; *1A*, first anal. (From Comstock and Needham)

B, front wing of an adult female of the narrow-winged tree cricket

C, front wing of an immature male, showing widening of inner half to form vibrating area, or tympanum, and modification of veins in this area. (From Comstock and Needham)

D, right front wing of adult male of the narrow-winged tree cricket; the second branch of cubitus (*Cu₂*) becomes the curved file vein (*fv*); *s*, the scraper

Some crickets have large wings, some small wings, some no wings at all. The females are provided with long ovipositors for placing their eggs in twigs of trees or in the ground (Figs. 35, 36).

The musical or stridulating organs of the crickets are similar to those of the katydids, being formed from the veins of the basal parts of the front wings. But in the crickets the organs are equally developed on each wing, and it looks as if these insects could play with either wing uppermost. Yet most of them consistently keep the *right* wing on top and use the file of this wing and the scraper of the left, just the reverse of the custom among the katydids.

The front wings of male crickets are usually very broad and have the outer edges turned down in a wide flap that folds over the sides of the body when the wings are closed. The wings of the females are simpler and usually smaller. The differences between the front wings in the male and the female of one of the tree crickets (Fig. 37) is shown at B and D of Figure 33. The inner half of the wing (or the rear half when the wing is extended) is very large in the male (D) and has only a few veins, which brace or stiffen the wide membranous vibratory area or *tympanum*. The inner basal part, or *anal area*, of the male wing is also larger than in the female and contains a prominent vein (Cu_2) which makes a sharp curve toward the edge of the wing. This vein has the stridulating file on its under surface. The veins in the wing of an adult female (B) are comparatively simple, and those of a young female (A) are more so. But the complicated venation of the male wing has been developed from the simple type of the female, which is that common to insects in general. The wing of a young male (C) is not so different from that of a young female (A) but that the corresponding veins can be identified, as shown by the lettering. Taking next the wing of the adult male (D), it is an easy matter to determine which veins have been distorted to produce the stridulating apparatus. When the tree crickets sing they elevate the wings above the back like two broad fans (Figs. 37, 40) and move them sidewise so that the file of the right rubs over the scraper of the left.

Fig. 34. A mole cricket, *Neocurtilla hexadactyla*

THE MOLE CRICKETS

The mole crickets (Fig. 34) are solemn creatures of the earth. They live like true moles in burrows underground, usually in wet fields or along streams. Their forefeet are broad and turned outward for digging like the front feet of moles. But the mole crickets differ from real moles in having wings, and sometimes they leave their burrows at night and fly about, being occasionally attracted to lights. Their front wings are short and lie flat on the back over the base of the abdomen, but the long hind wings are folded lengthwise over the back and project beyond the tip of the body.

Notwithstanding the gloomy nature of their habitat, the male mole crickets sing. Their music, however, is solemn and monotonous, being always a series of loud, deep-toned chirps, like *churp, churp, churp*, repeated very regularly about a hundred times a minute and continued indefinitely if the singer is not disturbed. Since the notes are most frequently heard coming from a marshy field or from the edge of a stream, they might be supposed to be those of a small frog. It is difficult to capture a mole cricket in the act of singing, for he is most likely standing at an opening in his burrow into which he retreats before he is discovered.

THE FIELD CRICKETS

This group of crickets includes *Gryllus* as its typical member, but entomologists give first place to a smaller brown cricket called *Nemobius*. There are numerous species of this genus, but a widely distributed one is *N. vittatus*, the striped ground cricket. This is a little cricket, about three-eighths of an inch in length, brownish in color, with three darker stripes on the abdomen, common in fields and dooryards (Fig. 35). In the fall the females lay their eggs in the ground with their slender ovipositors (D, E) and the eggs (F) hatch the following summer.

The song of the male Nemobius is a continuous twitter-

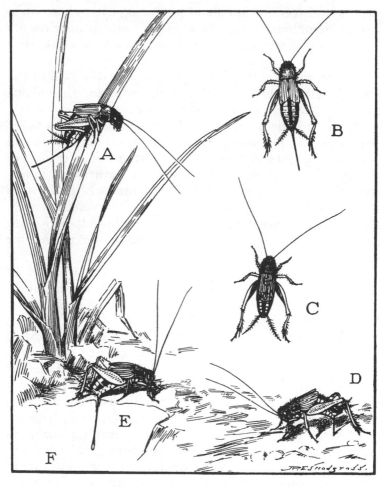

FIG. 35.   The striped ground cricket, *Nemobius vittatus*

A, B, females, distinguished by the long ovipositor.   C, a male.   D, a female
in the act of thrusting her ovipositor into the ground.   E, a female, with ovipositor full length in the ground, and extruding an egg from its tip.   F, an egg in
the ground

ing trill so faint that you must listen attentively to hear it.
In singing the male raises his wings at an angle of about
45°.   The stridulating vein is set with such fine ridges that

they would seem incapable of producing even those whispering Nemobius notes. Most of the muscial instruments of insects can be made to produce a swish, a creak, or a grating noise of some sort when handled with our clumsy fingers or with a pair of forceps, but only the skill of the living insect can bring from them the tones and the volume of sound they are capable of producing.

Our best-known cricket is Gryllus, the black cricket (Fig. 36), so common everywhere in fields and yards and occasionally entering houses. The true house cricket of Europe, *Gryllus domesticus*, has become naturalized in this country and occurs in small numbers through the Eastern States. But our common native species is *Gryllus assimilis*. Entomologists distinguish several varieties, though they are inclined to regard them all as belonging to the one species.

Mature individuals of Gryllus are particularly abundant in the fall; in southern New England they appear every year at this season by the millions, swarming everywhere, hopping across the country roads in such numbers that it is impossible to ride or walk without crushing them. Most of the females lay their eggs in September and October, depositing them singly in the ground (Fig. 36 D, E) in the same way that Nemobius does. These eggs hatch about the first of June the following year. But at this same time another group of individuals reaches maturity, a group that hatched in midsummer of the preceding year and passed the winter in an immature condition. The males of these begin singing at Washington during the last part of May, in Connecticut the first of June, and may be heard until the end of June. Then there is seldom any sound of Gryllus until the middle of August, when the males of the spring group begin to mature. From now on their notes become more and more common and by early fall they are to be heard almost continuously day and night until frost.

The notes of Gryllus are always vivacious, usually cheerful, sometimes angry in tone. They are merely chirps, and may be known from all others by a broken or vibratory sound. There is little music in them, but the player has enough conceit to make up for this lack. Two vigorous

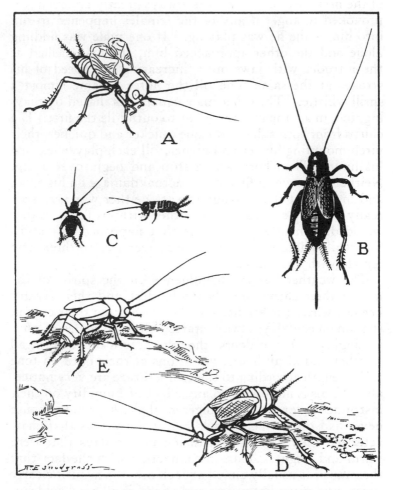

FIG. 36. The common black cricket, *Gryllus assimilis*

A, a male with wings raised in the attitude of singing. B, a female with long ovipositor. C, young crickets recently hatched (enlarged about 2½ times). D, a female inserting her ovipositor in the ground. E, a female with ovipositor buried full length in the ground

males that were kept in a cage together with several females gave each other little peace. Whenever one began to play his fiddle the other started up, to the plain disgust of the first one, and either was always greatly annoyed and provoked to anger if any of the females happened to run into him while he was playing. If one male was fiddling alone and the other approached him, the first dashed at the intruder with jaws open, increasing the speed of his strokes at the same time till the notes became almost a shrill whistle. The other male usually retaliated by playing, too, in an apparent attempt to outfiddle the first. The chirps from both sides now came quicker and quicker, their pitch mounting higher and higher, till each player reached his limit. Then both would stop and begin over again. Neither male ever inflicted any actual damage on his rival, and in spite of their savage threats neither was ever seen really to grasp any part of the other with his jaws. Either would dash madly at a female that happened to disturb him while fiddling, but neither was ever seen to threaten a female with open jaws.

The weather has much influence on the spirits of the males; their chirps are always loudest and their rivalry keenest when it is bright and warm. Setting their cage in the sun on cold days always started the two males at once to singing. Out of doors, though the crickets sing in all weather and at all hours, variations of their notes in tone and strength according to the temperature are very noticeable. This is not owing to any effect of humidity on their instruments, for the two belligerent males kept in the house never had the temper on cold and gloomy days that characterized their actions and their song on days that were warm and bright. This, in connection with the fact that their music is usually aimed at each other in a spirit clearly suggestive of vindictiveness and anger, is all good evidence that Gryllus sings to *express himself* and not to "charm the females." In fact, it is often hard to feel certain whether he is singing or swearing. If we could understand the words, we might be shocked at the awful language he is hurling at his rival. However, swearing is only a form of emotional expression, and singing is another. Gryllus, like an opera singer, simply expresses *all* his emotions in music, and, whether we can understand the words or not, we understand the sentiment.

At last one of the two caged rivals died; whether from natural causes or by foul means was never ascertained. He was alive early on the day of his demise but apparently weak, though still intact. In the middle of the afternoon, however, he lay on his back, his hind legs stretched out straight and stiff; only a few movements of the front legs showed that life was not yet quite extinct. One antenna was lacking and the upper lip and adjoining parts of the face were gone, evidently chewed off. But this is not necessarily evidence that death had followed violence, for, in cricketdom, violence more commonly follows death; that is, cannibalism is substituted for interment. A few days before, a dead female in the cage had been devoured quickly, all but the skull. After the death of this male, the remaining one no longer fiddled so often, nor with the same sharp challenging tone as before. Yet this could not be attributed to sadness; he had despised his rival and had clearly desired to be rid of him; his change was due rather to the lack of any special stimulus for expression.

THE TREE CRICKETS

The unceasing ringing that always rises on summer evenings as soon as the shadows begin to darken, that shrill melody of sound that seems to come from nothing but from everywhere out of doors, is mostly the chorus of the tree crickets, the blend of notes from innumerable harpists playing unseen in the darkness. This sound must be the most familiar of all insect sounds, but the musicians themselves are but little known to the general public. And when one of them happens to come to the window or into the house and plays in solo, the sound is so surprisingly

loud that the player is not suspected of being one of that
band whose mingled notes are heard outside softened by
distance and muffled by screens of foliage.

Out of doors the music of an individual cricket is so
elusive that even when you think you have located the ex-

FIG. 37. The snowy tree cricket, *Oecanthus niveus*
The upper figures, males, the one on the right with fore wings
raised vertically in attitude of singing; below, a female, with
narrow wings folded close against the body

act bush or vine from which it comes the notes seem to
shift and dodge. Surely, you think, the player must be
under that leaf; but when you approach your ear to it, the
sound as certainly comes from another over yonder; but
here you are equally convinced that it comes from still

another place farther off. Finally, though, it strikes the ear with such intensity that there can be no mistaking the source of its origin, and, right there in plain sight on a leaf sits a little, delicate, slim-legged, pale-green insect with hazy, transparent sails outspread above its back. But can such an insignificant creature be making such a deafening sound! It has required very cautious tactics to approach thus close without stopping the music, and it needs but a touch on stem or leaf to make it cease. But now those gauzy sails that before were a blurred vignette have acquired a definite outline, and a little more disturbance may cause them to be lowered and spread flat on the creature's back. The music will not begin anew until you have passed a period of silent waiting. Then, suddenly, the lacy films go up, once more their outlines blur, and that intense scream again pierces your ear. In short, you are witnessing a private performance of the broad-winged tree cricket, *Oecanthus latipennis*.

But if you pay attention to the notes of other singers, you will observe that there is a variety of airs in the medley going on. Many notes are long trills like the one just identified, lasting indefinitely; but others are softer purring sounds, about two seconds in length, while still others are short beats repeated regularly a hundred or more times every minute. The last are the notes of the snowy tree cricket, *Oecanthus niveus*, so-called on account of his paleness. He is really green in color, but a green of such a very pale shade that he looks almost white in the dark. The male (Fig. 37) is a little longer than half an inch, his wings are wide and flat, overlapping when folded on the back, with the edges turned down against the sides of the body. The female is heavier-bodied than the male, but her wings are narrow, and when folded are furled along the back. She has a long ovipositor for inserting her eggs into the bark of trees.

The males of the snowy cricket reach maturity and begin to sing about the middle of July. The singer raises his wings vertically above the back and vibrates them sidewise so rapidly that they are momentarily blurred with each note. The sound is that *treat, treat, treat, treat* already described, repeated regularly, rhythmically, and monotonously all through the night. At the first of the season there may be about 125 beats every minute, but later, on hot nights, the strokes become more rapid and mount to 160 a minute. In the fall again the rate decreases on cool evenings to perhaps a hundred. And finally, at the end of the season, when the players are benumbed with cold, the

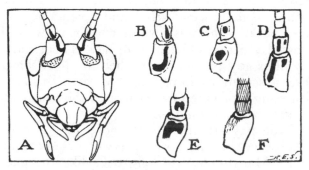

Fig. 38. Distinguishing marks on the basal segments of the antennae of common species of tree crickets

A, B, narrow-winged tree cricket, *Oecanthus angustipennis*. C, snowy tree cricket, *niveus*. D, four-spotted tree cricket, *nigricornis quadripunctatus*. E, black-horned tree cricket, *nigricornis*. F, broad-winged tree cricket, *latipennis*

notes become hoarse bleats repeated slowly and irregularly as if produced with pain and difficulty.

The several species of tree crickets belonging to the genus *Oecanthus* are similar in appearance, though the males differ somewhat in the width of the wings and some species are more or less diffused with a brownish color. But on their antennae most species bear distinctive marks (Fig. 38) by which they may be easily identified. The snowy cricket, for example, has a single oval spot of black on the under side of each of the two basal antennal joints (Fig. 38 C). Another, the narrow-winged tree cricket, has a spot on the second joint and a black J on the first (A, B). A third, the four-spotted cricket (D), has a dash and dot side by side on each joint.   A fourth, the black-horned or striped tree cricket (E), has two spots on each joint more or less run together, or sometimes has the whole base of the antenna blackish, while the color may also spread over the fore parts of the body and, on some individuals, form

Fig. 39.   Male and female of the narrow-winged tree cricket, *Oecanthus angusti-pennis*

The female is feeding on a liquid exuded from the back of the male, while the latter holds his fore wings in the attitude of singing.   (Enlarged about 3 times)

stripes along the back.   A fifth species, the broad-winged (F), has no marks on the antennae, which are uniformly brownish.

The narrow-winged tree cricket (*Oecanthus angusti-pennis*) is almost everywhere associated with the snowy, but its notes are very easily distinguished.   They consist of slower, purring sounds, usually prolonged about two seconds, and separated by intervals of the same length, but as fall approaches they become slower and longer.   Always they are sad in tone and sound far off.

The three other common tree crickets, the black-horned or striped cricket, *Oecanthus nigricornis*, the four-spotted,

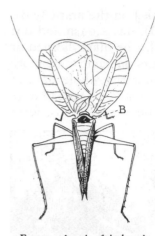

FIG. 40. A male of the broad-winged tree cricket, *Oecanthus latipennis*, with wings elevated in position of singing, seen from above and behind, showing the *basin* (*B*) on his back into which the liquid is exuded that attracts the female

O. nigricornis quadripunctatus, and the broad-winged, *O. latipennis*, are all trillers; that is, their music consists of a long, shrill whir kept up indefinitely. Of these the broad-winged cricket makes the loudest sound and the one predominant near Washington. The black-horned is the common triller farther north, and is particularly a daylight singer. In Connecticut his shrill note rings everywhere along the roadsides, on warm bright afternoons of September and October, as the player sits on leaf or twig fully exposed to the sun. At this season also, both the snowy and the narrow-winged sing by day but usually later in the afternoon and generally from more concealed places.

We should naturally like to know why these little creatures are such persistent singers and of what use their music is to them. Do the males really sing to charm and attract the females as is usually presumed? We do not know; but sometimes when a male is singing, a female approaches him from behind, noses about on his back, and soon finds there a deep basinlike cavity situated just behind the bases of the elevated wings. This basin contains a clear liquid which the female proceeds to lap up very eagerly,

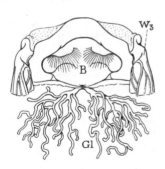

FIG. 41. The back of the third thoracic segment of the broad-winged tree cricket, with its basin (*B*) that receives secretion from the glands (*Gl*) inside the body

as the male remains quiet with wings upraised though he has ceased to play (Fig. 39). We must suspect, then, that in this case the female has been attracted to the male rather by his confectionery offering than by his music. The purpose of the latter, therefore, would appear to be to advertise to the female the whereabouts of the male, who she knows has sweets to offer; or if the liquid is sour or bitter it is all the same—the female likes it and comes after it. If, now, this luring of the female sometimes ends in marriage, we may see here the real reason for the male's possessing his music-making organs and his instinct to play them so continuously.

A male cricket with his front wings raised, seen from above and behind as he might look to a female, is shown in Figure 40. The basin (B) on his back is a deep cavity on the dorsal plate of the third thoracic segment. A pair of large branching glands (Fig. 41, *Gl*) within the body open just inside the rear lip of the basin, and these glands furnish the liquid that the female obtains.

There is another kind of tree cricket belonging to another genus, *Neoxabia*, called the two-spotted tree cricket, *N. bipunctata*, on account of two pairs of dark spots on the wings of the female. This cricket is larger than any of the species of *Oecanthus* and is of a pinkish brown color. It is widely distributed over the eastern half of the United States, but is comparatively rare and seldom met with. Allard says its notes are low, deep, mellow trills continued for a few seconds and separated by short intervals, as are the notes of the narrow-winged Oecanthus, but that their tone more resembles that of the broad-winged.

THE BUSH CRICKETS

The bush crickets differ from the other crickets in having the middle joint in the foot larger and shaped more like the third joint in the foot of a katydid (Fig. 17 B). Among the bush crickets there is one notable singer common in the neighborhood of Washington. This is the jumping bush cricket, *Orocharis saltator* (Fig. 42), who comes on the stage late in the season, about the middle of August, or shortly after. His notes are loud, clear, piping chirps with a rising inflection toward the end, suggestive of the notes of a small tree toad, and they at once strike the listener as

something new and different in the insect program. The players, however, are at first very hard to locate, for they do not perform continuously—one note seems to come from here, a second from over there, and a third from a different angle, so that it is almost impossible to place any one of them. But after a week or so the crickets become more numerous and each

FIG. 42. The jumping bush cricket, *Orocharis saltator*
Upper figure, a male; lower, a female

player more persistent till soon their notes are the predominant sounds in the nightly concerts, standing out loud and clear against the whole tree-cricket chorus. As Riley says, this chirp "is so distinctive that when once studied it is never lost amid the louder racket of the katydids and other night choristers."

After the first of September it is not hard to locate one of the performers, and when discovered with a flashlight, he is found to be a medium-sized, brown, short-legged cricket, built somewhat on the style of *Gryllus* but smaller (Fig. 42). The male, however, while singing raises his wings straight up, after the manner of the tree crickets, and he too, carries a basin of liquid on his back much sought after

[70]

by the female. In fact the liquid is so attractive to her that, at least in a cage, she is sometimes so persistent in her efforts to obtain it that the male is clearly annoyed and tries to avoid her. One male was observed to say very distinctly by his actions, as he repeatedly tried to escape the nibbling of a female, presumably his wife since she was taken with him when captured, "I do wish you would quit pestering me and let me sing!" Here is another piece of evidence suggesting that the male cricket sings to express his own emotions, whatever they may be, and not primarily to attract the female. But if, as in the case of the tree crickets, his music tells the female where she may find her favorite confection, and this in turn leads to matrimony, when the male is in the proper mood, it suggests a practical use and a reason for the stridulating apparatus and the song of the male insect.

WALKING-STICKS AND LEAF INSECTS

Talent often seems to run in families, or in related families, but it does not necessarily express itself in the same way. If the katydids and crickets are noted musicians, some of their relatives, belonging to the family

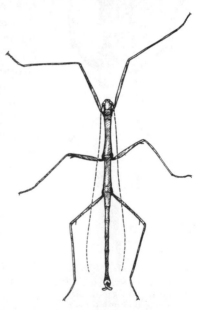

FIG. 43. The common walking-stick insect, *Diapheromera femorata*, of the eastern part of the United States. (Length 2½ inches)

Phasmidae, are incomparable mimics. Their mimicry, however, is not a conscious imitation, but is one bred in their bodily forms through a long line of ancestors.

If sometime in the woods you should chance to see a short, slender piece of twig suddenly come to life and slowly walk away on six slim legs, the marvel would not be a miracle, but a walking-stick insect (Fig. 43). These insects are fairly common in the eastern parts of the United States, but on account of their resemblance to twigs, and their habit of remaining perfectly quiet for a long time with the body pressed close to a branch of a tree, they are more frequently overlooked than seen. Sometimes, however, they occur locally in great numbers. It is supposed that the stick insects so closely resemble twigs for the purpose of protection from their enemies, but it has not been shown just what enemies they avoid by their elusive shape. The stick insects are more common in the South and in tropical countries, where some attain a remarkable length, one species from Africa, for example, being eleven inches long when full-grown. In New Guinea there lives a species that looks more like a small club than a stick, it being a large, heavy-bodied, spiny creature, nearly six inches in length and an inch in width through the thick-est part of its body (Fig. 44).

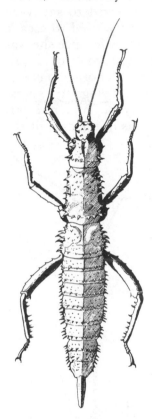

FIG. 44. A gigantic spiny walking-stick insect, *Eury-canthus horrida*, from New Guinea. (Length 5½ inches)

Other members of the phasmid family have specialized on imitating leaves. These insects have wings in the adult stage, and, of course, the wings make it easier for

[72]

them to take the form of leaves. One famous species that lives in the East Indies looks so much like two leaves stuck together that it is truly marvelous that an insect could be so fashioned (Fig. 45). The whole body is flat, and about three inches long, the bases of the legs are broad and irregularly notched, the abdomen is spread out almost as thin as a real leaf, and the leaflike wings are held close above it. Finally, the color, which is leaf-green or brown, gives the last touch necessary for complete dissimulation.

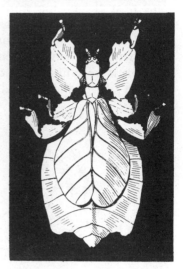

FIG. 45. A tropical leaf insect, *Pulchriphyllium pulchrifolium*, a member of the walking-stick family. (Length 3 inches)

THE MANTIDS

It is often observed that genius may be perverted, or put to evil purposes. Here is a family of insects, the Mantidae, related to the grasshoppers, katydids, and crickets, the members of which are clever enough, but are deceitful and malicious.

The praying mantis, *Stagmomantis carolina* (Fig. 46), though he may go by the aliases of "rear-horse" and "soothsayer," gets his more common name from the prayerful attitude he commonly assumes when at rest. The long, necklike prothorax, supporting the small head, is elevated and the front legs are meekly folded. But if you examine closely one of these folded legs, you will see that the second and third parts are armed with suspicious-looking spikes, which are concealed when the two parts are closed upon each other. In truth, the mantis is an arch hypocrite, and his devotional attitude and meek looks betoken no humility of spirit. The spiny arms, so innocently folded upon the breast, are direful weapons held ready to strike as soon as some unsuspecting insect happens within their reach. Let a small grasshopper come near the posing saint: immediately a sly tilt of the head belies the suppliant manner, the crafty eyes leer upon the approaching insect, losing no detail of his movements. Then, suddenly, without warning, the praying mantis becomes a demon in action. With a nice calculation of distance, a swift movement, a snatch of the

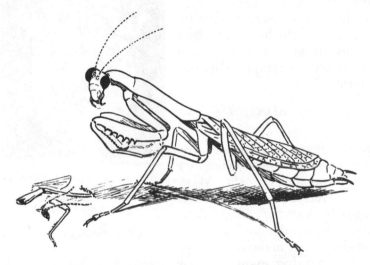

FIG. 46. The praying mantis, *Stagmomantis carolina*, and remains of its last meal. (Length 2½ inches)

terrible clasps, the unlucky grasshopper is a doomed captive, as securely held as if a steel trap had closed upon his body. As the hapless creature kicks and wrestles, the jaws of the captor sink into the back of his head, evidently in search of the brain; and hardly do his weakening struggles cease before the victim is devoured. Legs, wings, and other fragments unsuitable to the taste of an epicure are thrown aside, when once more the mantis sinks into repose, piously folds his arms, and meekly awaits the

[74]

chance arrival of the next course in his ever unfinished banquet of living fare.

Some exotic species of mantids have the sides of the prothorax extended to form a wide shield (Fig. 47), beneath which the forelegs are folded and completely hidden. It is not clear what advantage they derive from this device, but it seems to be one more expression of deceit.

Of course, as we shall take occasion to observe later, goodness and bad-ness are largely matters of relativity.

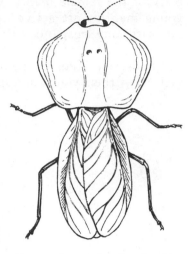

Fig. 47.   A mantis from Ecuador with a shieldlike extension of its back. (Length 3⅜ inches)

Fig. 48.   Egg case of a mantis attached to a twig,   *Stagmomantis carolina*

The mantis is an evil creature from the standpoint of a grasshopper, but he would be regarded as a benefactor by those who have a grudge against grass-hoppers or against other insects that the mantis destroys. Hence, we must reckon the mantis as at least a beneficial insect relative to human welfare. A large species of mantis, introduced a few years ago into the eastern States from China, is now regarded as a valuable agricul-tural asset because of the number of harmful insects it destroys.

The mantids lay their eggs in large cases stuck to the twigs of trees (Fig. 48). The substance of which the case is made is similar to that with which the locusts inclose their eggs, and is exuded from the body of the female mantis when the eggs are laid. The young mantids are active little creatures, without wings but with long legs, and it is the fate of those unprotected green bugs, the aphids, or plant lice, that infest the leaves of almost all kinds of plants, to become the principal victims of their youthful appetites.

CHAPTER III

ROACHES AND OTHER ANCIENT INSECTS

WE used to speak quite confidently of time as something definite, measurable by the clock, and of a year or a century as specific quantities of duration. In this present age of relativity, however, we do not feel so certain about these things. Geologists calculate in years the probable age of the earth, and the length of time that has elapsed since certain events took place upon it, but their figures mean only that the earth has gone around the sun approximately so many times during the interval. In biology it signifies nothing that one animal has been on the earth for a million years, and another for a hundred million, for the unit of evolution is not a year, but a generation. If one animal, such as most insects, has from one to many generations every year, and another, such as man, has only four or five in a century, it is evident that the first, by evolutionary reckoning, will be vastly older than the second, even though the two have made the same number of trips with the earth around the sun. An insect that antedates man by several hundred million years, therefore, is ancient indeed.

The roach scarcely needs an introduction, being quite well known to all classes of society in every inhabited part of the world. That he has long been established in human communities is shown by the fact that the various nations have bestowed different names upon him. His common English name of "cockroach" is said to come from the Spanish, *cucaracha*. The Germans call him, rather disrespectfully, *Küchenschabe*, which signifies "kitchen

[77]

louse." The ancient Romans called him *Blatta*, and on this his scientific family name of Blattidae is based. A small species of Europe, named by the entomologists

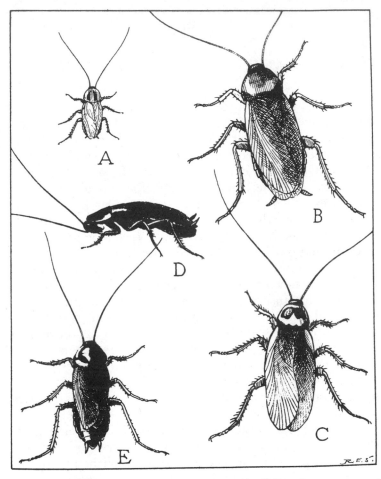

FIG. 49. The four species of common household roaches

A, the German roach, or Croton bug, *Blattella germanica* (length ⅝ inch). B, the American cockroach, *Periplaneta americana* (length 1⅜ inches). C, the Australian cockroach, *Periplaneta australasiae* (length 1¼ inches). D, the wingless female of the Oriental roach, *Blatta orientalis* (length 1⅛ inches). E, the winged male of the Oriental roach (length 1 inch)

Blattella germanica, which is now our most common American roach, received the nickname of "Croton bug" in New York, because somehow he seemed to spread with the introduction of the Croton Valley water system, and this appelation has stuck to him in many parts of the country.

The Croton bug, or German roach (Fig. 49 A), is the smallest of the "domestic" varieties of roaches. It is that rather slender, pale-brown species, about five-eighths of an inch in length, with the two dark spots on the front shield of its body. This roach is the principal pest of the kitchen in the eastern part of the United States, and prob-

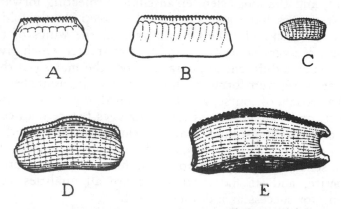

FIG. 50. Egg cases of five species of roaches. (Twice natural size)
A, egg case of the Australian roach (fig. 49 C). B, that of the American roach (fig. 49 B); the other three are made by out-of-door species

ably the best support of the trade in roach powders. Several other larger species are fortunately less numerous, but still familiar enough. Among these are one called the American roach (Fig. 49 B), a second known as the Australian roach (C), and a third as the Oriental roach (D, E). These four species of cockroaches are all great travelers and recognize no ties of nationality. They are equally at home on land and at sea, and, as uninvited passengers on ships, they have spread to all countries where ships have gone.

Besides the household roaches, there are great numbers of species that live out of doors, especially in warm and tropical regions. Most of these are plain brown of various shades, or blackish, but some are green, and a few are spotted, banded, or striped. Different species vary much in size, some of the largest reaching a length of four inches, measured to the tips of the folded wings, while the smallest are no longer than three thirty-seconds of an inch in length. They nearly all have the familiar flattened form, with the head bent down beneath the front part of the body, and the long, slender antennae projecting forward. Most species have wings which they keep closely folded over the back. In the Oriental roach, the wings of the female are very short (Fig. 49 D), a character which gives them such a different appearance from the males (E) that the two sexes were formerly supposed to be different species.

The roach, of course, was not designed to be a household insect, and it lived out of doors for ages before man constructed dwellings, but it happens that its instincts and its form of body particularly adapt it to a life in houses. Its keen sense, its agility, its nocturnal habits, its omnivorous appetite, and its flattened shape are all qualities very fitting for success as a domestic pest.

Many kinds of roaches give birth to living young; but most of our common species lay eggs, which they inclose in hard-shelled capsules. The material of the capsule is a tough but flexible substance resembling horn, and is produced as a secretion by a special gland in the body of the female opening into the egg duct. The capsule is formed in the egg duct, and the eggs are discharged into it while the case is held in the orifice of the duct. When the receptacle is full its open edge is closed, and the eggs are thus tightly sealed within it. The sealed border is finely notched, and transverse impressions on the surface of the capsule indicate the position of the eggs within it.

The Croton bug, or German roach (Fig. 49 A), makes a small flat tabloid egg case, which the female usually carries about with her for some time projecting from the end of her body, and sometimes the eggs hatch while she is still carrying the case.   The American and Australian roaches (Fig. 49 B, C) make egg cases much resembling miniature pocketbooks or tobacco pouches, about three-eighths or half an inch in length, with a serrated clasp along the upper edge (Fig. 50 A, B).   The cases of some of the smaller species of roaches are only one-sixteenth of an inch long

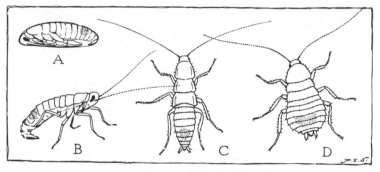

Fig. 51. Young of the German roach, or Croton bug (fig. 49 A), in various stages just before and after hatching

A, the young roach in the egg just before hatching.   B, the young roach just after hatching, shedding its embryonic covering membrane.   C, young roach after shedding the embryonic covering.   D, the same individual half an hour old

(C), while larger species may make a case three-quarters of an inch in length (E).

The embryo roaches mature within the eggs, and when they are ready to hatch they emerge inside the egg case. By some means, the roughened edge of the case where it was last closed is opened to allow the imprisoned insects to escape.   Small masses of the tiny creatures now bulge out, and finally the whole wriggling contents of the capsule is projecting from the slit.   First one or two individuals free themselves, then several together fall out, then more of them, until soon the case containing the empty eggshells is deserted.

When the young roaches first liberate themselves from the capsule, they are helpless creatures, for each is contained in a close-fitting membrane that binds its folded legs and antennae tightly to the body and keeps the head pressed down against the breast (Fig. 51 A). The inclosing sheath, however, a film so delicate as to be almost invisible, is soon burst by the struggling of the little roach anxious to be free—it splits and rapidly slides down over the body (B), from which it is at last pushed off. The shrunken, discarded remnant of the skin is now such an insignificant flake that it scarce seems possible it so recently could have enveloped the body of the insect.

The newly liberated young roach dashes off on its slim legs with an activity quite surprising in a creature that has never had the use of its legs before. It is so slender of figure (Fig. 51 C) that it does not look like a roach, and it is pale and colorless except for a mass of bright green material in its abdomen. But, almost at once, it begins to change; the back plates of the thorax flatten out, the body shortens by the overlapping of its segments, the abdomen takes on a broad, pear-shaped outline, the head is retracted beneath the prothoracic shield, and by the end of half an hour the little insect is unmistakably a young cockroach (D).

The roaches have a potent enemy in the house centipede, that creature of so many legs (Fig. 52) that it looks like an animated blur as it occasionally darts across the living-room floor or disappears in the shades of the basement before you are sure whether you have seen something or not, but which is often trapped in the bathtub, where its appearance is likely to drive the housewife into hysteria. Unless you are fond of roaches, however, the house centipede should be protected and encouraged. The writer once placed one of these centipedes in a covered glass dish containing a female Croton bug and a capsule of her eggs which were hatching. No sooner were the young roaches running about than the centipede began a feast which ended only when the last of the brood had been devoured. The mother roach was not at the time molested, but next morning she lay dead on her back, her head severed and dragged some distance from the body, which was sucked dry of its juices—mute evidence of the tragedy that had befallen some-time in the night, probably when the pangs of returning hunger stirred the centipede to renewed activity. The house centipede does not confine itself to a diet of live roaches, for it will eat almost any kind of food, but it is never a pest of the household larder.

Most species of roaches have two pairs of well-developed wings, which they ordinarily keep folded over the back, for in their usual pursuits the domestic spe-cies do not often fly, except oc-casionally when hard pressed to avoid capture. The front wings are longer and thicker than the hind wings, and are laid over the latter, which are thin and folded fanwise when not in use. In these characters the roaches resemble the grasshoppers and katydids,

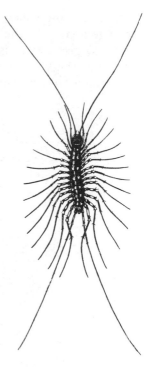

Fig. 52. The common house centipede, *Scutigera forceps* (natural size), a destroyer of young roaches

and their family, the Blattidae, is usually placed with these insects in the order Orthoptera.

The wings of insects are interesting objects to study. When spread out flat, as are those of the roach shown in Figure 53, they are seen to consist of a thin membranous tissue strengthened by many branching ribs, or *veins*, extending outward from the base. The wings of all insects are constructed on the same general plan and have the same primary veins; but, since the great specialty of insects is flight, in their evolution they have concentrated on the wings, and the different groups have tried out different styles of venation, with the result that now each is distinguished by some particular pattern in the arrangement of the veins and their branches. The entomologist can thus not only distinguish by their wing structure the various orders of insects, as the Orthoptera, the dragonflies, the moths, the bees, and the flies, but in

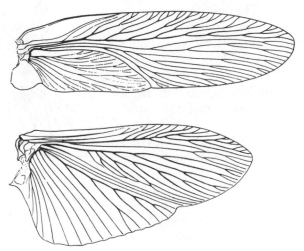

Fig. 53. Wings of a cockroach, *Periplaneta*, showing the vein pattern characteristic of the roach family

many cases he can identify families and even genera. Particularly are the wings of value to the student of fossil insects, for the bodies are so poorly preserved in most cases that without the wings the paleontologist could have made little headway in the study of insects of the past. As it is, however, much is known of insects of former times, and a study of their fossil remains has contributed a great deal to our knowledge of this most versatile and widespread group of animals.

ROACHES AND OTHER ANCIENT INSECTS

The paleontological history of life on the earth shows us that the land has been inhabited successively by different forms of animals and plants. A particular group of creatures appears upon the scene, first in comparative insignificance; then it increases in numbers, in diversity of forms, and usually in the size of individuals, and may become the dominant form of life; then again it falls back to insignificance as its individuals decrease in size, its species in numbers, until perhaps its type becomes extinct. Meanwhile another group, representing another type of structure, comes into prominence, flourishes, and declines. It is a mistake, however, to get the impression that all forms of life have had this succession of up and down in their history, for there are many animals that have existed with little change for immense periods of time.

The history of insects gives us a good example of permanence. The insects must have begun to be insects somewhere in those remote periods of time before the earliest known records of animals were preserved in the rocks. They must have been present during the age when the water swarmed with sharks and great armored fishes; they certainly flourished during the era when our coal beds were being deposited; they saw the rise of the huge amphibians and the great reptilian beasts, the *Dinosaurus*, the *Ichthyosaurus*, the *Plesiosaurus*, the *Mosasaurus*, and all the rest of that monster tribe whose names are now familiar household words and whose bones are to be seen in all our museums. The insects were branching out into new forms during the time when birds had teeth and were being evolved from their reptile ancestors, and when the flowering plants were beginning to decorate the landscape; they were present from the beginning of the age of mammals to its culmination in the great fur-bearing creatures but recently extinct; they attended the advent of man and have followed man's whole evolution to the present time; they are with us yet—a vigorous race that shows no sign of weakening or of decrease in numbers. Of all the land animals, the insects are the true blue-blood aristocrats by length of pedigree.

The first remains of insects known are found in the upper beds of the rocks laid down in the geological period of the earth's history known as the Carboniferous. Dur-

FIG. 54. A group of common Carboniferous plants reaching the size and proportions of large trees. (From Chamberlin and Salisbury, drawn by Mildred Marvin from restorations of fossil specimens.) Courtesy of Henry Holt & Co.

Of the two large trees in the foreground, the one on the left is a *Sigillaria*, that on the right a *Lepidodendron;* of the two large central trees in the background the left is a *Cordaites*, the right a tree fern; the tall stalks in the outermost circle are *Calamites*, plants related to our horsetail ferns

ing Carboniferous times much of the land along the shores of inland seas or lakes was marshy and supported great forests from which our coal deposits have been formed. But the Carboniferous landscape would have had a strange and curious look to us, accustomed as we are to an abundance of hard-wood, leafy trees and shrubs, and a multitude of flowering plants. None of these forms of vegetation had yet appeared.

Much of the undergrowth of the Carboniferous swamps was composed of fernlike plants, many of which were, indeed, true ferns, and perhaps the ancestors of our modern brackens. Some of these ancient ferns grew to a great size, and rose above the rest in treelike forms, attaining a height of sixty feet and more, to branch out in a feathery crown of huge spreading fronds. Another group of plants characteristic of the Carboniferous flora comprised the seed ferns, so named because, while closely resembling ferns in general appearance, they differed from true ferns in that they bore seeds instead of spores. The seed ferns were mostly small plants with delicate, ornate leaves, and they have left no descendants to modern times.

Along with the numerous ferns and seed ferns in the Carboniferous swamps, there were gigantic club mosses, or lycopods, which, ascending to a height sometimes of much more than a hundred feet, were the conspicuous big trees in the forests of their day (Fig. 54). These lycopods had long, cylindrical trunks covered with small scales arranged in regular spiral rows. Some had thick branching limbs starting from the upper part of the trunk and closely beset with stiff, sharp-pointed leaves; others bore at the top of the trunk a great cluster of long slender leaves, giving them somewhat the aspect of a gigantic variety of our present-day yucca, or Spanish bayonet. The bases of the larger trees expanded to a diameter of three or four feet, and were supported on huge spreading underground branches from which issued the roots—a device, perhaps, that gave them an ample foundation in the soft mud of the swamps in which they grew.

The Carboniferous lycopods furnished most of our coal, and then, in later times, their places were taken by other types of vegetation. But their race is not yet extinct, for we have numerous representatives of them with us today in those lowly evergreen plants known as club mosses, whose spreading, much-branched limbs, usually trailing on the ground, are covered by rows of short, stiff leaves. The most familiar of the club mosses, though not a typical species, is the "ground pine." This humble little shrub, so much sought for Christmas decoration, still in some places carpets our woods with its soft, broad, frondlike stems. In the fall when its rich dark green so pleasingly contrasts with the somber tones of the season's dying foliage, it seems to be an expression of the vitality that has preserved the lycopod race through the millions of years which have elapsed since the days of its great ancestors. The "resurrection plant," often sold to housekeepers under false or exaggerated claims of a marvelous capacity for rejuvenation, is also a descendant of the proud lycopods of ancient times.

In our present woodlands, along the banks of streams or in other moist places, there grows also another plant that has been preserved to us from the Carboniferous forests—the common "horsetail fern," or *Equisetum*, that green, rough-ribbed stalk with the whorls of slender branches growing from its joints. Our equisetums are modest plants, seldom attaining a height of more than a few feet, though in South American countries some species may reach an altitude of thirty feet; but in Carboniferous times their ancestors grew to the stature of trees (Fig. 54) and measured their robust stalks with the trunks of the lycopods and giant ferns.

Aside from the numerous representatives of these several groups of plants, all more or less allied to the ferns, the Carboniferous forests contained another group of treelike plants, called *Cordaites*, from which the cycads of later times and our present-day maidenhair tree, or ginko, are probably descended. Then, too, there were a few representatives of a type that gave origin to our modern conifers.

ROACHES AND OTHER ANCIENT INSECTS

It is probable that a visitor to those days of long ago might give us a more complete account of the vegetation that grew in the Carboniferous swamps than can be known from the records of the rocks, but the paleobotanist has a wealth of material now at hand sufficient to give us at least a pretty reliable picture of the setting in which the earliest of known insects lived and died.

And now, what were the insects like that inhabited the forests of those early times? Were they, too, strangely fashioned creatures, fit denizens of a far-off fairyland? No, nothing of the sort, at least not in appearance or structure, though "fit" they probably were, from a physical standpoint, for insects are fitted to live almost anywhere. In short, the Carboniferous insects were principally roaches! Yes, those woods and swamps of millions of years ago were alive with roaches little different from our own familiar household pests, or from the numerous species that have not forsaken their native habitats for life in the cities.

Whoever looks to the geological records for evidence of the evolution of insects is sorely disappointed, for even in the venation of the wings those early roaches (Fig. 55) were almost identical with our present species (Fig. 53). As typical examples of the Carboniferous roaches, the species shown in Figure 55 serve well, and anyone can see, even though the specimens lack antennae and legs, that the creatures were just common roaches. Hence, we can easily picture these ancient roaches scuttling up the tall trunks of the scaly lycopods, and shuffling in and out among the bases of the close-set leaf stems of the tree ferns, and we should expect to find an abundant infestation of them in the vegetational refuse matted on the ground. Insects of those days must have been comparatively free from enemies, for birds did not yet exist, and all that host of parasitic insects that attack other insects were not evolved until more recent times.

Though by far the greater number of the Carboniferous

insects known are roaches, or insects closely related to roaches, there were many other forms besides. Some of these are of particular interest to entomologists because, in some ways, they are more simple in structure than are

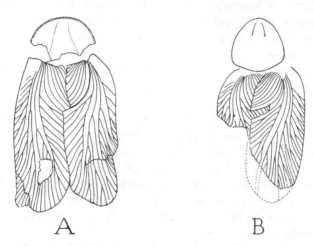

FIG. 55. Fossil cockroaches from Upper Carboniferous rocks A, *Asemoblatta mazona*, found in Illinois, length of wing one inch. (From Handlirsch after Scudder.) B, *Phyloblatta carbonaria*, found in Germany. (From Handlirsch)

any of the modern insects, and in this respect they apparently stand closer to the hypothetical primitive insects than do any others that we know. And yet, the characters by which these oldest known insects, called the *Paleodictyoptera*, differ from modern forms are so slight that they would scarcely be noticed by anyone except an entomologist; to the casual observer, the Paleodictyoptera would be just insects. Their chief distinguishing marks are in the pattern of the wing venation, which is more symmetrical than in other winged insects, and, therefore, probably closer to that of the primitive ancestors of all the winged insects. These ancient insects probably did not fold the wings over the back, as do most present-day insects, showing thus another primitive character, though not a distinctive one, since modern dragonflies (Fig. 58) and mayflies (Fig. 60) likewise keep the wings extended when at rest.

The question of how insects acquired wings is always one of special interest, since, while we know perfectly well that the wing of a bird or of a bat is merely a modified fore limb, the nature of the primitive organ from which the insect wing has been evolved is still a mystery. The Paleodictyoptera, however, may throw light upon the subject, for some of them had small flat lobes on the lateral edges of the back plate of the prothorax, which in fossil specimens look like undeveloped wings (Fig. 56). The presence of these prothoracic lobes, occurring as they do in some of the oldest known insects, has suggested the

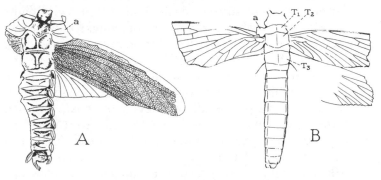

Fig. 56. Examples of the earliest known fossil insects, called the Paleodictyoptera, having small lobes (*a*) projecting like wings from the prothorax

A, *Stenodictya lobata* (from Brongniart).   B, *Eubleptus danielsi* (drawn from specimen in U. S. Nat. Mus.): T_1, T_2, T_3, back plates of three thoracic segments

idea that the true wings were evolved from similar flaps of the mesothorax and metathorax. If so, we must picture the immediate ancestors of the winged insects as creatures provided with a row of three flaps on each side of the body projecting stiffly outward from the edges of the thoracic segments. Of course, the creatures could not actually fly with wings of this sort, but probably they could glide through the air from the branches of one tree to another as well as can a modern flying squirrel by means of the folds of skin stretched along the sides of its body between the fore and the hind legs. If such lobes then became flexible at their bases, it required only a slight adjustment of the muscles already present in the body to give them motion in an up-and-down direction; and the wings of modern insects, in most cases, are still moved by a very simple mechanism which has involved the acquisition of few extra muscles.

It appears, however, that three pairs of fully-developed wings would be too many for mechanical efficiency. In the later evolution of insects, therefore, the prothoracic lobes were never developed beyond the glider stage, and in all modern insects this first pair of lobes has been lost. Furthermore, it was subsequently found that swift flight is best attained with a *single* pair of wings; and nearly all the more perfected insects of the present time have the hind pair of wings reduced in size and locked to the front pair to insure unity of action. The flies have carried this evolution toward a two-winged condition so far that they have practically achieved the goal, for with them the hind wings are so greatly reduced that they no longer have the form or function of organs of flight, and these insects, named the Diptera, or two-winged insects, fly with one highly specialized and efficient pair of wings (Fig. 167).

The Paleodictyoptera became extinct by the end of the Carboniferous period, and their disappearance gives added support to the idea that they were the last survivors of an earlier type of insect. But they were by no means the primitive ancestors of insects, for, in the possession of wings alone, they show that they must have undergone a long evolution while wings were in the course of development; but of this stage in the history of insects we know nothing. The rocks, so far as has yet been revealed, contain no records of insect life below the upper beds of the Carboniferous deposits, when insects were already fully winged. This fact shows how cautious we must be in making negative statements concerning the extinct inhabitants of the earth, for we know that insects must have lived long before we have evidence of their existence. The absence of insect fossils earlier than the Carboniferous is hard to explain, because for millions of years the remains of other animals and plants had

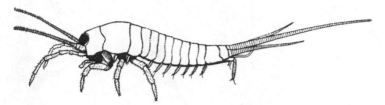

Fig. 57. *Machilis*, a modern representative of ancient insects before the development of wings. (Length of body ⅜ inch)

been preserved, and have since been found in compara-tive abundance. As a consequence, we have no concrete knowledge of insects before they became winged creatures evolved almost to their modern form.

At the present time there are wingless insects. Some of them show clearly that they are recent descendants from winged forms. Others suggest by their structure that their ancestors never had wings. Such as these, therefore, may have come down to us by a long line of descent from the primitive wingless ancestors of all the insects. The common "fish moth," known to entomolo-gists as *Lepisma*, and its near relation, *Machilis* (Fig. 57), are familiar examples of the truly wingless insects of the present time, and if their remote ancestors were as fragile and as easily crushed as they, we may see a reason why they never left their impressions in the rocks.

Along with the Carboniferous roaches and the Paleo-dictyoptera, there lived a few other kinds of insects, many of which are representative of certain modern

FIG. 58. Dragonflies (Order Odonata), modern representatives of an ancient group of winged insects. The adults are strong fliers,

groups. Among the latter were dragonflies, and some of these must have been of gigantic size, for insects, because they attained a wing expanse of fully two feet, while the largest of modern dragonflies do not measure more than eight inches across the expanded wings. But the length of wing of the extinct giant dragonflies does not necessarily mean that the bulk of the body was much greater than that of the largest insects living today. In general, the insects of the past were of ordinary size, the majority of them probably matching with insects of the present time.

The modern dragonflies (Fig. 58) are noted for their rapid flight and for the ability to make instantaneous changes in the direction of their course while flying. These qualities enable them to catch other insects on the wing, which constitute their food. Their wings are provided with sets of special muscles, such as other insects do not possess, showing that the dragonflies are descended along a line of their own from their Carboniferous progenitors. They still retain a character of their ancestors in that they are unable to fold the wings flat over the back in the manner that most other insects fold their wings when they are not using them. The larger dragonflies hold the wings straight out from the sides of the body when at rest (Fig. 58); but a group of slender dragonflies, known as the damselflies (Plate 1, Fig. 2), bring the wings together over the back in a vertical plane.

The dragonflies are usually found most abundantly in the neighborhood of open bodies of water. Over the unobstructed surface of the water the larger species find a convenient hunting ground; but a more important reason for their association with water is that they lay their eggs either in the water or in the stems of plants growing in or beside it. The young dragonflies (Fig. 59) are aquatic and must have an easy access to water. They are homely, often positively ugly, creatures, having none of the elegance of their parents. They feed on other living creatures which their swimming powers enable them to pursue, and which they capture by means of grasping hooks on the end of their extraordinarily long underlip (Fig. 134 A), which can be shot out in front of the head (B). The great swampy lakes of Paleozoic times must have furnished an ideal habitat for dragonflies, and it is probable that the most ancient dragonflies known had a structure and habits not very different from those of modern species.

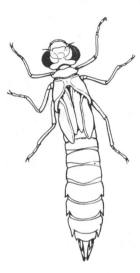

Another very common insect of the present time, which appears likewise to be a direct descendant of Paleozoic ancestors, is the mayfly (Fig. 60). The young mayflies (Fig. 61) also live in the water, and are provided with gills for aquatic breathing, having the form of flaps or filaments situated in a row along each side of the body. The adults (Fig. 60) are very delicate insects with four gauzy wings, and a pair of long threadlike tails projecting from the rear end of the body. At the time of their transformation they often issue in great swarms from the water, and they are particularly attracted to strong lights.

FIG. 59. A young dragon-fly, an aquatic creature that leaves the water only when ready to transform into the adult (fig. 58)

For this reason large numbers of them come to the cities at night, and in the morning they may be seen sitting about on walls and windows, where they find themselves in a situation totally strange to their native habits and instincts. The mayflies do not fold their wings horizontally, but when at rest bring them together vertically over the back (Fig. 60). In this respect they, too, appear to preserve a character of their Paleozoic ancestors; though it must be observed that the highly evolved modern butterflies close their wings in the same fashion.

ROACHES AND OTHER ANCIENT INSECTS

The roaches, the dragonflies, and the mayflies attest the great antiquity of insects, for since these forms existed practically as they are today in Paleozoic times, the primitive ancestors of all the insects, of which we have no remains in the geological records, must have lived in times vastly more remote. However, though we may search in vain the paleontological records for evidence of the origin and early development of insects, the subsequent evolution of the higher forms of modern insects is clearly shown by the species preserved in eras later

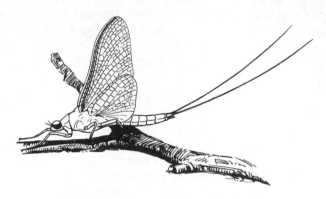

Fig. 60. A mayfly, representative of another order of primitive winged insects having numerous relations in Paleozoic times. (Twice natural size)

than the Carboniferous. Such insects as the beetles, the moths, the butterflies, the wasps, the bees, and the flies are entirely absent in the older rocks, but make their appearance at later periods or in comparatively recent times, thus confirming the idea derived from a study of their structure that they have been evolved from ancestors more closely resembling the paleodictyopteran types of the Carboniferous beds.

The long line of descent of the roach, with almost no change of form or structure, furnishes material for a special lesson in evolution. If evolution has been a matter of survival of the fittest, the roach, judged by survival, must be a most fit insect. Its fitness, however, is of a general nature; it is one that adapts the roach to live successfully in many kinds of conditions and circum-

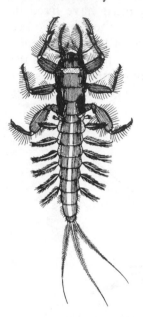

stances. Most other forms of modern insects have been evolved through an adaptation to more special kinds of habitats and to particular ways of living or of feeding. Such insects we say are *specialized*, while those exemplified in the roach are said to be *generalized*. Survival, therefore, may depend either on generalization or on specialization. Generalized forms of animals have a better chance of surviving through a series of changing conditions than has an animal which is specifically adapted to one kind of life, though the latter may have an advantage as long as conditions are favorable to it.

FIG. 61. A young mayfly, a water-inhabiting creature. (One-half larger than natural size)

The roaches, therefore, have survived to present times, and will probably live as long as the earth is habitable, because, when driven from one environment, they make themselves at home in another; but we have all seen how the specialized mosquito disappears when its breeding places are destroyed. From this consideration we can draw some consolation for the human race, if we do not mind likening ourselves to roaches; for, as the roach, man is a versatile animal, capable of adapting himself to all conditions of living, and of thriving in extremes.

CHAPTER IV

WAYS AND MEANS OF LIVING

In our human society each individual must obtain the things necessary for existence; the manner by which he acquires them, whether by one trade or another, by this means or by that, does not physically matter so long as he provides himself and his family with food, clothing, and shelter. Exactly so it is with all forms of life. The physical demands of living matter make certain things necessary for the maintenance of life in that matter, but nature has no law specifying that any necessity shall be acquired in a certain manner. Life itself is a circumscribed thing, but it has complete freedom of choice in the ways and means of living.

It is useless to attempt to make a definition of what living matter is, or of how it differs from non-living matter, for all definitions have failed to distinguish animate from non-animate substance. But we all know that living things are distinguishable from ordinary non-living things by the fact that they make some kind of response to changes in the contact between themselves and their environment. The "environment," of course, must be broadly interpreted. Biologically, it includes all things and forces that in any way touch upon living matter. Not only has every plant and animal as a whole its environment, but every part of it has an environment. The cells of an animal's stomach, for example, have their environment in the blood and lymph on one side, the contents of the stomach on the other; in the energy of the nerves distributed to them; and in the effects of heat and cold that penetrate them.

The environmental conditions of the life of cells in a complex animal are too complicated for an elemental study; the elements of life and its basic necessities are better understood in a simple organism, or in a one-celled animal; but for purposes of description, it is most convenient to speak of the properties of mere *protoplasm.* All the vital needs of the most highly organized animal are present in any part of the protoplasmic substance of which it is composed.

Protoplasm is a chemical substance, or group of substances, the structure of which is very complex but is maintained so long as there is no disturbance in the environ-

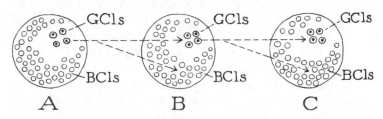

FIG. 62.   Diagram showing the relation of the germ cells (*GCls*) and the body cells (*BCls*) in successive generations

A fertilized germ cell of generation A forms the germ cells and body cells of B, a fertilized germ cell of B forms the germ cells and body cells of C, and so on. The offspring C of B derives nothing from the body cells of the parent B, but both offspring C and parent B have a common origin in a germ cell of A

ment.   Let some least thing happen, however, such as a change in the temperature, in the strength of the light, in the weight of pressure, or in the chemical composition of the surrounding medium, and the protoplasmic molecules, in the presence of oxygen, are likely to have the balance of their constituent particles upset, whereupon they partly decompose by the union of their less stable elements with oxygen to form simpler and more permanent compounds. The decomposition of the protoplasmic substances, like all processes of decomposition, liberates a certain amount of energy that had been stored in the making of the molecule, and this energy may manifest itself in various ways.   If it takes the form of a change of shape in the protoplasmic mass, or movement, we say the mass exhibits signs of life. The state of being alive, however, is more truly shown if the act can be repeated, for the essential property of living matter is its power of reverting to its former chemical composition, and its ability thus gained of again reacting to another change in the environment. In restoring its lost elements, it must get these elements anew from the environment, for it can not take them back from the substances that have been lost.

Here, expressed in its lowest terms, is the riddle of the physical basis of life and of the incentive to evolution in the forms of life. Not that these mysteries are any more easily understood for being thus analyzed, but they are more nearly comprehended. Being alive is maintaining the power of repeating an action; it involves sensitivity to stimuli, the constant presence of free oxygen, elimination of waste, and a supply of substances from which carbon, hydrogen, nitrogen, and oxygen, or other necessary elements, are readily available for replacement purposes. Evolution results from the continual effort of living matter to perform its life processes in a more efficient manner, and the different groups of living things are the result of the different methods that life has tried and found advantageous for accomplishing its ends. Living organisms are machines that have become more and more complex in structure, but always for doing the same things.

If animals may be compared with machines in their physical mechanism, they are like them, too, in the fact that they wear out and are at last beyond repair. But here the simile ends, for when your car will no longer run, you must go to the dealer and order a new one. Nature provides continuous service by a much better scheme, for each organism is responsible for its own successor. This phase of life, the replacement of individuals, opens another subject involving ways and means, and it, likewise, can be understood best in its simpler manifestations.

INSECTS

The facts of reproduction in animals are not well expressed by our name for them. Instead of "reproduction," it would be truer to say "repeated production," for individuals do not literally reproduce themselves. Generations are serially related, not each to the preceding; they follow one another as do the buds along the twig of a tree,

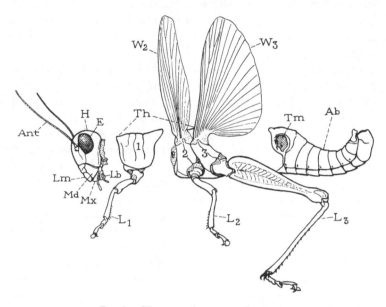

FIG. 63. The external structure of an insect

The body of a grasshopper dissected showing the head (*H*), the thorax (*Th*), and the abdomen (*Ab*). The head carries the eyes (*E*), the antennae (*Ant*), and the mouth parts, which include the labrum (*Lm*), the mandibles (*Md*), the maxillae (*Mx*), and the labium (*Lb*). The thorax consists of three segments (*1, 2, 3*), the first separate and carrying the first legs (*L₁*), the other two combined and carrying the wings (*W₂, W₃*), and the second and third legs (*L₂, L₃*). The abdomen consists of a series of segments; that of the grasshopper has a large tympanal organ (*Tm*), probably an ear, on each side of its base. The end of the abdomen carries the external organs of reproduction and egg-laying

and buds on the same twig are identical or nearly so, not because one produces the next, but because all are the result of the same generative forces in the twig. If the spaces of the twig between the buds were shortened until one bud became contiguous with the one before, or became enveloped by it, a relation would be established between the two buds similar to that which exists between successive generations of life forms.   The so-called parent generation, in other words, *contains* the germs of the succeeding generation, but it does not *produce* them. Each generation is simply the custodian of the germ cells entrusted to it, and the "offspring" resembles the parent, *not* because it is a chip off the parental block, but because both parent and offspring are developed from the same line of germ cells.

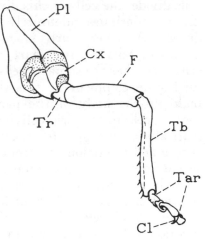

FIG. 64.   The leg of a young grasshopper, showing the typical segmentation of an insect's leg

The leg is supported on a pleural plate (*Pl*) in the lateral wall of its segment. The basal segment of the free part of the leg is the coxa (*Cx*), then comes a small trochanter (*Tr*), next a long femur (*F*) separated by the knee bend from the tibia (*Tb*), and lastly the foot, consisting of a sub-segmented tarsus (*Tar*), and a pair of terminal claws (*Cl*) with an adhesive lobe between them

Parents create the conditions under which the germ cells will develop; they nourish and protect them during the period of their development; and, when each generation has served the purpose of its existence, it sooner or later dies.   But the individuals produced from its germ cells do the same for another set of germ cells produced simultaneously with themselves, and so on as long as the species persists.

To express the facts of succession in each specific form of animal, then, we should analyze each generation into *germ cells* and an accompanying mass of protective cells which

[103]

forms a body, or *soma*, the so-called parent. Both the body, or somatic, cells and the germ cells are formed from a single primary cell, which, of course, is usually produced by the union of two incomplete germ cells, a spermatozoon and an egg. The primary germ cell divides, the daughter cells divide, the cells of this division again divide, and the division continues indefinitely until a mass of cells is produced. At a very early stage of division, however, two groups of cells are set apart, one representing the germ cells, the other the somatic cells. The former refrain from further development at this time; the latter proceed to build up the body of the parent. The relation of the somatic cells to the germ cells may be represented diagrammatically as in Figure 62, except that the usual dual parentage and the union of germ cells is not expressed. The sexual form of reproduction is not necessary with all lower animals, nor with all generations of plants; in some insects the eggs can develop without fertilization.

The fully-developed mass of somatic cells, whose real function is that of a servant to the germ cells, has assumed such an importance, as public servants are prone to do, that we ordinarily think of it, the body, the active sentient animal, as the essential thing. This attitude on our part is natural, for we, ourselves, are highly organized masses of somatic cells. From a cosmic standpoint, however, no creature is important. Species of animals and plants exist because they have found ways and means of living that have allowed them to survive, but the physical universe cares nothing about them—the sunshine is not made for them, the winds are not tempered to suit their convenience. Life must accept what it finds and make the best of it, and the question of how best to further its own welfare is the problem that confronts every species.

The sciences of anatomy and physiology are a study of the methods by which the soma, or body, has contrived to meet the requirements imposed upon it by the unchanging laws of the physical universe. The methods adopted are as

numerous as the species of plants and animals that have
existed since life began. A treatise on entomology, therefore, is an account of the ways and means of living that
insects have adopted and perfected in their somatic organization. Before discussing insects in particular, however,
we must understand a little more fully the principal conditions of living that na-
ture places on all forms of
life.

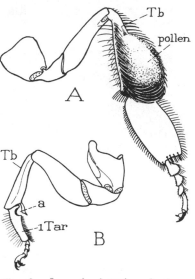

As we have seen, life is
a series of chemical re-
actions in a particular
kind of matter that can
carry on these reactions.
A "reaction" is an action;
and every act of living
matter involves a break-
ing down of some of the
substances in the proto-
plasm, the discharging of
the waste materials, and
the acquisition of new
materials to replace those
lost. The reaction is in-
herent in the physical or

Fig. 65. Legs of a honeybee, showing
special modifications

A, outer surface of a hind leg, with a
pollen basket on the tibia (*Tb*) loaded
with pollen. B, a fore leg, showing the
antenna cleaner (*a*) between the tibia and
the tarsus, and the long, hairy basal
segment of the tarsus (*1 Tar*), which is
used as a brush for cleaning the body

chemical properties of
protoplasmic compounds
and depends upon the
substances with which
the protoplasm is sur-
rounded. It is the func-
tion of the creature's mechanism to see that the conditions surrounding its living cells are right for the continuance of the cell reactions. Each cell must be
provided with the means of eliminating waste material
and of restoring its lost material, since it can not utilize
that which it has discarded.

With the conditions of living granted, however, protoplasm is still only potentially alive, for there is yet required a *stimulus* to set it into activity. The stimulus for life activities comes from changes in the physical forms of energy that surround or infringe upon the potentially living substance; for, "live" matter, like all other matter, is subject to the law of inertia, which decrees that it must remain at rest until motion is imparted to it by other

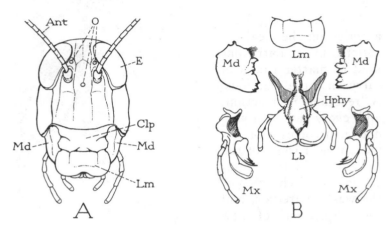

FIG. 66.   The head and mouth parts of a grasshopper

A, facial view of the head, showing the positions of the antennae (*Ant*), the large compound eyes (*E*), the simple eyes, or ocelli (*O*), the broad front lip, or labrum (*Lm*) suspended from the cranium by the clypeus (*Clp*), and the bases of the mandibles (*Md*, *Md*) closed behind the labrum

B, the mouth parts separated from the head in relative positions, seen from in front: *Hphy*, hypopharynx, or tongue, attached to base of labium; *Lb*, labium; *Lm*, labrum; *Md*, mandibles; *Mx*, maxillae

motion.   A very small degree of stimulating energy, however, may result in the release of a great quantity of stored energy.

The food of all living matter must contain carbon, hydrogen, nitrogen, and oxygen.   The mechanism of plants enables them to take these elements from compounds dissolved in the water of the soil.   Animals must get them from other living things, or from the products of living things. Therefore, animals principally have developed the power of movement; they have acquired grasping organs of some sort, a mouth, and an alimentary canal for holding the food when once obtained.

In the insects, the locomotory function is subserved by the legs and by the wings. Since all these organs, the three pairs of legs and the two pairs of wings, are carried by the thorax (Fig. 63, *Th*), this region of the body is distinctly the locomotor center of the insect. The legs (Fig. 64) are adapted, by modifications of structure in different species, for walking, running, leaping, digging, climbing, swimming, and for many varieties of each of these ways of progression, fitting each species for its particular mode of living and of obtaining its food. The wings of insects are important accessions to their locomotory equipment, since they greatly increase their means of getting about, and thereby extend their range of feeding. The legs, furthermore, are often modified in special ways to perform some function accessory to feeding. The honeybee, as is well known, has pollen-collecting brushes on its front legs (Fig. 65 B), and pollen-carrying baskets on its hind legs (A). The mantis, which captures other insects and eats them alive, has its front legs made over into those efficient organs for grasping its prey and for holding the struggling victim which have already been described (Fig. 46).

The principal organs by which insects obtain and manipulate their food consist of a set of appendages situated on the head in the neighborhood of the mouth, which, in their essential structure, are of the nature of the legs, for insects have no jaws comparable with those of vertebrate animals. The mouth appendages, or *mouth parts* as they are called, are very different in form in the various groups of insects that have different feeding habits, but in all cases they consist of the same fundamental pieces. Most important is a pair of jawlike appendages, known as the *mandibles* (Fig. 66 B, *Md*), placed at the sides of the mouth (A, *Md*), where they swing sidewise and close upon each other below the mouth. Behind the mandibles is a pair of *maxillae* (B, *Mx*) of more complicated form, fitted rather for holding the food than for crushing it. Following the maxillae is a large under lip, or *labium* (*Lb*), having the

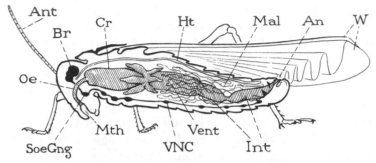

FIG. 67. Lengthwise section of a grasshopper, showing the general location of the principal internal organs, except the respiratory tracheal system and the organs of reproduction

An, anus; *Ant*, antenna; *Br*, brain; *Cr*, crop; *Ht*, heart; *Int*, intestine; *Mal*, Malpighian tubules; *Mth*, mouth; *Oe*, oesophagus; *SoeGng*, suboesophageal ganglion; *Vent*, stomach (ventriculus); *VNC*, ventral nerve cord; *W*, wings

structure of two maxillae united by their inner margins. A broad flap hangs downward before the mouth to form an upper lip, or *labrum* (*Lm*). Between the mouth appendages and attached to the front of the labium there is a large median lobe of the lower head wall behind the mouth, known as the *hypopharynx* (*Hphy*).

Insects feed, some on solid foods, others on liquids, and their mouth parts are modified accordingly. So it comes about that, according to their feeding habits, insects may be separated into two groups, which, like the fox and the stork, could not feed either at the table of the other. Those insects, such as the grasshoppers, the crickets, the beetles, and the caterpillars, that bite off pieces of food tissue and chew them, have the mandibles and the other mouth parts of the type described above. Insects that partake only of liquids, as do the plant lice, the cicadas, the moths, the butterflies, the mosquitoes and other flies, have the mouth parts fitted for sucking, or for piercing and sucking. Some of the sucking types of mouth parts will be described in other chapters (Figs. 121, 163, 183), but it will be seen that all are merely adaptations of form based on the ordinary biting type of mouth appendages. The fossil records of the history of insects show that the sucking insects are the more recent products of evolution, since all the earlier kinds of insects, the cockroaches and their kin, have typical biting mouth parts.

The principal thing to observe concerning the organs of feeding, in a study of the physiological aspect of anatomy, is that they serve in all cases to pass the natural food materials from the outside of the animal into the alimentary canal, and to give them whatever crushing or mastication is necessary. It is within the alimentary canal, therefore, that the next steps toward the final nutrition of the animal take place.

The alimentary canal of most insects is a simple tube (Fig. 68), extending either straight through the body, or

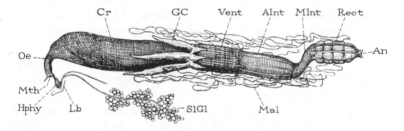

FIG. 68. The alimentary canal of a grasshopper

AInt, anterior intestine; *An*, anus; *Cr*, crop; *GC*, gastric caeca, pouches of the stomach; *Hphy*, hypopharynx (tongue); *Lb*, base of labium; *Mal*, Malpighian tubules; *MInt*, mid-intestine; *Mth*, mouth; *Oe*, oesophagus; *Rect*, hind intestine (rectum); *SlGl*, salivary glands opening by their united ducts at base of hypopharynx; *Vent*, ventriculus (stomach)

making only a few turns or loops in its course. It consists of three principal parts, of which the middle part is the true *stomach*, or *ventriculus* (*Vent*) as it is called by insect anatomists. The first part of the tube includes a *pharynx* immediately behind the mouth, followed by a narrower, tubular *oesophagus* (*Oe*), after which comes a sac-like enlargement, or *crop* (*Cr*), in which the food is temporarily stored, and finally an antechamber to the stomach, named the *proventriculus*. The third part of the alimentary canal, connecting the stomach with the anal opening, is the *intestine*, usually composed of a narrow anterior part, and a wide posterior part, or *rectum* (*Rect*). Muscle layers surrounding the entire alimentary tube cause the food to be swallowed and to be passed along from one section to the next toward the rear exit.

With the taking of the food into the alimentary canal, the matter of nutrition is by no means accomplished, for the animal is still confronted with the problem of getting the nutrient materials into the inside of its body, where alone they can be used. The alimentary tube has no openings anywhere along its course into the body cavity. Whatever food substances the tissues of the animal receive, therefore, must be taken *through* the walls of the tube in which they are inclosed, and this transposition is accomplished by dissolving them in a liquid. Most of the nutrient materials in the raw food matter, however, are not soluble in ordinary liquids; they must be changed chemically into a form that will dissolve. The process of getting the nutrient parts of the raw foodstuff into solution constitutes *digestion*.

The digestive liquids in insects are furnished mostly by the stomach walls or the walls of tubular glands that open into the stomach, but the secretion of a pair of large glands, called the *salivary glands* (Fig. 68, *SlGl*), which open between the mouth parts, perhaps has in some cases a digestive action on the food as it is taken into the mouth.

Digestion is a purely chemical process, but it must be a rapid one. Consequently the digestive juices contain not only substances that will transform the food materials into soluble compounds, but other substances that will speed up these reactions, for otherwise the animal would starve on a full stomach by reason of the slowness of its gastric service. The quickening substances of the digestive fluids are called *enzymes*, and each kind of enzyme acts on only one class of food material. An animal's practical digestive powers, therefore, depend entirely upon the specific enzymes its digestive liquids contain. Lacking this or that enzyme it can not digest the things that depend upon it, and usually its instincts are correlated with its enzymes so that it does not fill its stomach with food it can not digest. A few analyses of the digestive liquids of insects have been made, enough to show that their digestive processes depend upon the presence of the same enzymes as those of other animals, including man.

The grosser digestive substances, in cooperation with the enzymes, soon change all the parts of the food materials in the stomach that the animal needs for its sustenance into soluble compounds which are dissolved in the liquid part of the digestive secretions. Thus is produced a rich, nutrient juice within the alimentary canal which can be *absorbed* through the walls of the stomach and intestine and can so enter the closed cavity of the body. The next problem is that of *distribution*, for still the food materials must reach the individual cells of the tissues that compose the animal.

The insect's way of feeding, of digesting its food, and of absorbing it is not essentially different from that of the higher animals, including ourselves, for alimentation is a very old and fundamental function of all animals. Its means of distributing the digested food within its body, however, is quite different from that of vertebrates. The absorbed pabulum, instead of being received into a set of lymphatic vessels and from these sent into blood-filled tubes to be pumped to all parts of the organism, goes directly from the alimentary walls into the general body cavity, which is filled with a liquid that bathes the inner surfaces of all the body tissues. This body liquid is called the "blood" of the insect, but it is a colorless or slightly yellow-tinted lymph. It is kept in motion, however, by a pulsating vessel, or heart, lying in the dorsal part of the body; and by this means the food, now dissolved in the body liquid, is carried into the spaces between the various organs, where the cells of the latter can have access to it.

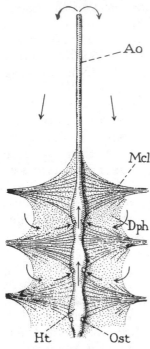

FIG. 69. Diagram of the typical structure of an insect's heart and supporting diaphragm, with the course of the circulating blood marked by arrows

Ao, aorta, or anterior tubular part of the heart without lateral openings; *Dph*, membranous diaphragm; *Ht*, anterior three chambers of the heart, which usually extends to the posterior end of the body; *Mcl*, muscles of diaphragm, the fibers spreading from the body wall to the heart; *Ost*, ostium, or one of the lateral openings into the heart chambers

The heart of the insect is a slender tube suspended along the midline of the back close to the dorsal wall of the body (Fig. 67, *Ht*). It has intake apertures along its sides (Fig. 69, *Ost*), and its anterior end opens into the body cavity. It pulsates forward, by means of muscle fibers in its walls, thereby sucking the blood in through the lateral openings and discharging it by way of the front exit. An imperfect circulation of the blood is thus established through the spaces between the organs of the body cavity, sufficient for the purposes of so small an animal as an insect.

The final act of nutrition comes now when the blood, charged with the nutrient materials absorbed from the digested food in the alimentary canal, brings these materials into contact with the inner tissues. The tissue cells, by the inherent power of all living matter (which depends on the laws of osmosis and on chemical affinity), take for themselves whatever they need from the menu offered by the blood, and with this matter they build up their own substance. It is evident, therefore, that the blood must contain a sufficient quantity and variety of dietary elements to satisfy all possible cell appetites; that the stomach's walls and their associated glands must furnish the enzymes appropriate for making the necessary elements available from the raw food matter in the stomach; and, finally, that it must be a part of the instincts of each animal species to consume such native foodstuffs of its environment as will supply every variety of nourishing elements that the cells demand.

As we have seen, the demand for food comes from the loss of materials that are decomposed in the tissues during cell activity. Better stated, perhaps, the chemical breakdown within the cell is the cause of the cell activity, or is the cell activity itself. The way in which the activity is expressed does not matter; whether by the contraction of a muscle cell, the secretion of a gland cell, the generation of nerve energy by a nerve cell, or just the minimum activity that maintains life, the result is the same always—the loss of certain substances. But, as with most chemical reduction processes, the protoplasmic activity depends upon the presence of available oxygen; for the decomposition of the unstable substances of the protoplasm is the result of the affinity of some of their elements for oxygen. Consequently, when the stimulus for action comes over a nerve from a nerve center, a sudden reorganization takes place between these protoplasmic elements and the oxygen atoms which results in the formation of water, carbon dioxide, and various stable nitrogenous compounds.

The substances discarded as a result of the cell activities are waste products, and must be eliminated from the organism for their presence would clog the further activity of the cells or would be poisonous to them. The animal, therefore, must have, in addition to its mechanisms for bringing food and oxygen to the cells, a means for the removal of wastes.

The supplying of oxygen and the removing of carbon dioxide and some of the excess water are accomplished by *respiration*. Respiration is primarily the exchange of gases between the cells of the body and the outside air. If an animal is sufficiently small and soft-skinned, the gas exchange can be made directly by diffusion through the skin. Larger animals, however, must have a device for conveying air into the body where the tissues will have closer access to it. It will be evident, then, that there is not necessarily only one way of accomplishing the purposes of respiration.

Vertebrate animals inhale air into a sac or pair of sacs, called the lungs, through the very thin walls of which the oxygen and carbon dioxide can go into and out of the blood respectively. The blood contains a special oxygen carrier in the red matter, hemoglobin, of its red corpuscles, by means of which the oxygen taken in from the air is transported to the tissues. The carbon dioxide is carried from the tissues partly by the hemoglobin, and partly dissolved in the blood liquid.

Insects have no lungs, nor have they hemoglobin in their blood, which, as we have seen, is merely the liquid that fills the spaces of the body cavity between the organs. Insects have adopted and perfected a method of getting air distributed through their bodies quite different from that of the vertebrates. They have a system of air tubes, called *tracheae* (Fig. 70), opening from the exterior by small breathing pores, or *spiracles* (*Sp*), along the sides of the body, and branching minutely within the body to all parts of the tissues. By this means the air is conveyed directly to the parts where respiration takes place. There are usually in insects ten pairs of spiracles, two on the sides of the thorax, and eight on the abdomen. The spiracles communicate with a pair of large tracheal trunks lying along the sides of the body (Fig. 70), and from these trunks are given off branches into each body segment and into the head, which go to the alimentary canal, the heart, the nervous system, the muscles, and to all the other organs, where they break up into finer branches that terminate in minute end tubes going practically to every cell of the body.

Many insects breathe by regular movements of expansion and contraction of the under surface of the abdomen, but experimenters have not yet agreed as to whether the air goes in and out of the same spiracles or whether it enters one set and is expelled through another. It is probable that the fresh air goes into the smaller tracheal branches principally by gas diffusion, for some insects make no perceptible respiratory movements.

The actual exchange of oxygen from the air and carbon dioxide from the tissues takes place through the thin walls of the minute end tubes of the tracheae. Since these tubes lie in immediate contact with the cell surfaces the gases do not have to go far in order to reach their destinations, and the insect has little need of an oxygen carrier in its blood—its whole body, practically, is a lung. And yet some investigations have made it appear likely that the insect blood does contain an oxygen carrier that functions in a manner similar to that of the hemoglobin of vertebrate blood, though the importance of oxygen transportation in insect physiology has

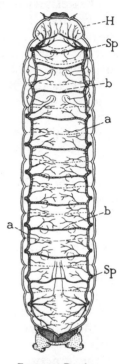

Fig. 70. Respiratory system of a caterpillar. The external breathing apertures, or spiracles (*Sp*, *Sp*), along the sides of the body open into lateral tracheal trunks (*a*, *a*), which are connected crosswise by transverse tubes (*b*, *b*) and give off minutely branching tracheae into all parts of the head (*H*) and body

not been determined. In any case, the tracheal method of respiration must be a very efficient one; for, considering the activity of insects, especially the rate at which the wing muscles act during flight, the consumption of oxygen must at times be pretty high.

The activity of insects depends very much, as every one knows, upon the temperature. We have all observed how the house flies disappear upon the first cold snap in the fall and then surprise us by showing up again when the weather turns warm, just after we have taken down the screens. All insects depend largely upon external warmth for the heat necessary to maintain cellular activity. While their movements produce heat, they have no means of conserving this heat in their bodies, as have "warm-blooded" animals. That insects radiate heat, however, is very evident from the high temperature that bees can maintain in their hives during winter by motion of the wings. All insects exhale much water vapor from their spiracles, another evidence of the production of heat in their bodies.

The solid matter thrown off from the cells in activity is discharged into the blood. These waste materials, which are mostly compounds of nitrogen in the form of salts, must then be removed from the blood, for their accumulation in the body would be injurious to the tissues. In vertebrate animals, the nitrogenous wastes are eliminated by the kidneys. Insects have a set of tubes, comparable with the kidneys in function, which open into the intestine at the junction of the latter with the stomach (Fig. 68, *Mal*), and which are named, after their discoverer, the *Malpighian tubules*. These tubes extend through the principal spaces of the body cavity, where they are looped and tangled like threads about the other organs and are continually bathed in the blood. The cells of the tube walls pick out the nitrogenous wastes from the blood and discharge them into the intestine, whence they are passed to the exterior with the undigested food refuse.

We thus see that the inside of an insect is not an unorganized mass of pulp, as believed by those people whose education in such matters comes principally from underfoot. The physical unity of all forms of life makes it necessary that every creature must perform the same vital functions. The insects have, in many respects, adopted their own ways of accomplishing these functions, but, as already pointed out, the means of doing a thing does not count with nature so long as the end results are attained. The essential conditions are the supply of necessities and the removal of wastes.

The body of a complex animal may be likened to a great factory, in which the individual workers are represented by the cells, and groups of workers by the organs. That the factory may accomplish its purpose, the activities of each worker must be coordinated with those of all the other workers by orders from a directing office. Just so, the activities of the cells and organs of the animal must be controlled and coordinated; and the directing office of the animal organization is the central nervous system. The work of almost every cell in the body is ordered and controlled by a "nerve impulse" sent to it over a nerve fiber from a nerve center.

The inner structure of the nervous tissues and the working mechanism of the nerve centers are essentially alike in all animals, but the form and arrangement of the nerve tissue masses and the distribution of the nerve fibers may differ much according to the plan of the general body organization. The insects, instead of following the vertebrate plan of having the central nerve cord along the back inclosed in a bony sheath, have found it just as well for their purposes to have the principal nerve cord lying free in the lower part of the body (Fig. 67, *VNC*). In the head there is a brain (Figs. 67, 72, *Br*) situated above the oesophagus (Fig. 67, *Oe*), but it is connected by a pair of cords with another nerve mass below the pharynx in the lower part of the head (*SoeGng*). From this nerve mass another pair of nerve cords goes to a third nerve mass

[117]

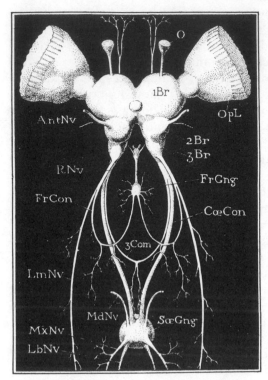

FIG. 71. The nervous system of the head of a grasshopper, as seen by removal of the facial wall

AntNv, antennal nerve; *1Br*, *2Br*, *3Br*, the three parts
of the brain; *CoeCon*, circumoesophageal connectives;
3Com, suboesophageal commissure of the third lobes
of the brain; *FrGng*, frontal ganglion; *FrCon*, frontal
ganglion connective with the brain; *LbNv*, labial
nerve; *LmNv*, labral nerve; *MdNv*, mandibular nerve;
MxNv, maxillary nerve; *O*, simple eye; *OpL*, optic
lobe connected with the brain; *RNv*, recurrent nerve;
SoeGng, suboesophageal ganglion

lying against the lower wall of the first body segment (Fig. 72, *Gng* 1), which is likewise connected with a fourth mass in the second segment, and so on. The central nervous system of the insect thus consists of a series of small nerve masses united by double nerve cords. The nerve masses are known as *ganglia* (*Gng*), and the uniting cords are called the *connectives* (Fig. 71, *Con*). Typically there is a ganglion for each of the first eleven body segments, besides the brain and the lower ganglion of the head.

The brain of an insect (Fig. 71) has a highly complex internal structure, but it is a less important controlling center than is the brain of a vertebrate animal. The other ganglia have much independence of function, each giving the stimuli for movements of its own segment. For this

[118]

reason, the head of an insect may be cut off and the rest of the creature may still be able to walk and to do various other things until it dies of starvation.   Similarly, with some species, the abdomen may be severed and the insect will still eat, though the food runs out of the cut end of the alimentary canal.   The detached abdomen may lay eggs, if properly stimulated.   Though the insect thus appears to be largely a creature of automatic regulations, acts are not initiated without the brain, and full coordination of the functions is possible only when the entire nervous system is intact.

The active elements of the nerve centers are nerve cells; the nerve fibers are merely conducting threads extended from the cells. If the nerve force that stimulates the other kinds of cells into activity comes from nerve cells, the question then arises as to whence comes the primary stimulus that activates the nerve cells.   We must discard the old idea that nerve cells act automatically; being matter, they are subject to the laws of matter—they are inert until compelled to act.   The stimulus of the nerve cells comes from something outside of them, either from the environmental forces of the external world or from substances formed by other cells within the body.

Nothing is known definitely of the internal stimuli of insects, but there can be no doubt that substances are formed by the physiological activities of the insect tissues, similar to the *hormones*, or secretions of the ductless glands of other animals, that control action in other organs either directly or through the nervous system. Thus, some internal condition must prompt the insect to feed when its stomach is empty, and the entrance of food into its pharynx must stimulate the alimentary glands to prepare the digestive juices. Probably a secretion from the reproductive organs of the female, when the eggs are ripe in the ovaries, gives the stimulus for mating, and later sets into motion the reflexes that govern the laying of the eggs. The caterpillar spins its cocoon at the proper time for doing so; the

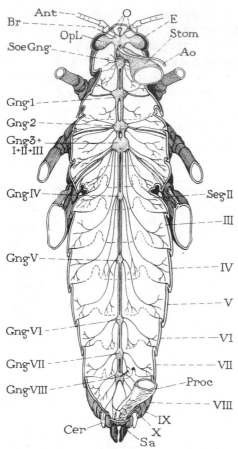

FIG. 72. The general nervous system of a grass-hopper, as seen from above

Ant, antenna; *Ao*, aorta; *Br*, brain; *Cer*, cercus; *E*, compound eye; *Gng1*, ganglion of prothorax; *Gng2*, ganglion of mesothorax; *Gng3+I+II+III*, compound ganglion of metathorax, comprising the ganglia belonging to the metathorax and the first three abdominal segments; *GngIV—GngVIII*, ganglia of the fourth to eighth abdominal segments; *O*, ocelli; *Proc*, proctodeum, or posterior part of alimentary canal; *Sa*, suranal plate; *SegII—X*, second to tenth segments of abdomen; *SoeGng*, suboesophageal ganglion; *Stom*, stomodeum, or anterior part of alimentary canal

stimulus, most likely, comes from the products of physiological changes beginning to take place in the body that will soon result in the transformation of the caterpillar into a chrysalis, a stage when the insect needs the protection of a cocoon. These activities of insects we call *instincts*, but the term is simply a cover for our ignorance of the processes that cause them.

External stimuli are things of the outer environment that affect the living organism. They include matter, electromagnetic energy, and gravity; but the known stimuli do not comprise all the activities of matter or of the "ether." The common stimuli are: pressure of solids, liquids, and gases; humidity; chemical qualities (odors and tastes);

sound, heat, light, and gravity. Most of these things stimulate the nerve centers indirectly through nerves connected with the skin or with specialized parts of the skin called *sense organs*. An animal can respond, therefore, only to those stimuli, or to the degrees of a particular stimulus, to which it is sensitive. If, for example, an animal has no receptive apparatus for sound waves, it will not be affected by sound; if it is not sensitized to certain wave lengths of light, the corresponding colors will not stimulate it. There are few *kinds* of natural activities in the environment that animals do not perceive; but even our own perceptive powers fall far short of registering all the *degrees* of any activity that are known to exist and which the physicist can measure.

Insects respond to most of the kinds of stimuli that we perceive by our senses; but if we say that they see, hear, smell, taste, or touch we make the implication that insects have consciousness. It is most likely that their reactions to external stimuli are for the most part performed unconsciously, and that their behavior under the effect of a stimulus is an automatic action entirely comparable to our reflex actions. Behavioristic acts that result from reflexes the biologist calls *tropisms*. Coordinated groups of tropisms constitute an instinct, though, as we have seen, an instinct may depend also on internal stimuli. It can not be said that consciousness does not play a small part in determining the activities of some insects, especially of those species in which memory, *i.e.*, stored impressions, appears to give a power of choice between different conditions presented. The subject of insect psychology, however, is too intricate to be discussed here.

The phases of life thus far described, the complexity of physical organization, the response to stimuli, the phenomena of consciousness from their lowest to their highest manifestations, all pertain to the soma. Yet, somehow, the plan of the edifice is carried along in the germ cells, and by them the whole somatic structure is rebuilt with but little change of detail from generation to generation. This phase of life activity is still a mystery to us, for no attempted explanation seems adequate to account for the organizing power resident in the germ cells that accomplishes the familiar facts of repeated

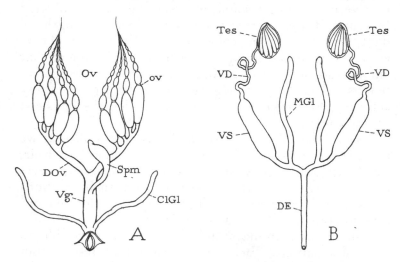

Fig. 73. Diagrams of the internal organs of reproduction in insects

A, the female organs, comprising a pair of ovaries (*Ov*), each composed of a group of egg tubules (*ov*), a pair of oviducts (*DOv*), and a median outlet tube, or vagina (*Vg*), with usually a pair of colleterial glands (*ClGl*) discharging into the vagina, and a sperm receptacle, or spermatheca (*Spm*), opening from the upper surface of the latter

B, the male organs, comprising a pair of testes (*Tes*) composed of spermatic tubules, a pair of sperm ducts, or vasa deferentia (*VD*), a pair of sperm vesicles (*VS*), and an outlet tube, or ductus ejaculatorius (*DE*), with usually a pair of mucous glands (*MGl*) discharging into the ducts of the sperm vesicles

development which we call reproduction. When we can explain the repetition of buds along the twig, we may have a key to the secret of the germ cells—and possibly to that of organic evolution.

The organs that house the germ cells in the mature insect consist of a pair of ovaries in the female (Fig. 73 A, *Ov*) in which the eggs mature, and of a pair of testes in the male (B, *Tes*) in which the spermatozoa reach their complete growth. Appropriate ducts connect the ovaries or the testes with the exterior near the rear end of the body. The female usually has a sac connected with the egg duct (A, *Spm*) in which the sperm, received at mating, are stored until the eggs are ready to be laid, when they are

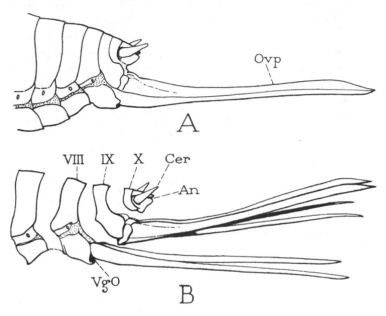

FIG. 74. The ovipositor of a long-horned grasshopper, a member of the katydid family, showing the typical structure of the egg-laying organ of female insects

A, the ovipositor (*Ovp*) in natural condition, projecting from near the posterior end of the body

B, the parts of the ovipositor separated, showing the six component pieces, two arising from the eighth abdominal segment (*VIII*), and four from the ninth (*IX*). *An*, anus; *Cer*, cerci; *IX*, ninth abdominal segment; *Ovp*, ovipositor; *VgO*, vaginal opening; *VIII*, eighth abdominal segment; *X*, tenth abdominal segment

extruded upon the latter and bring about fertilization. The egg cells ordinarily are all alike, but the spermatozoa are of two kinds; and according to the kind of sperm received by any particular egg, the future individual will be male or female.

INSECTS

The germ cells accompanying each new soma undergo a series of transformations within the parent body before they themselves are capable of accomplishing their purpose. They multiply enormously. With some animals, only a few of them ever produce new members of the race; but with insects, whose motto is "safety in numbers," each species produces every season a great abundance of new individuals, to the end that the many forces arrayed against them may not bring about their extermination.

The world seems full of forces opposed to organized life. But the truth is, all organization is an opposition to established forces. The reason that the forms of life now existing have held their places in nature is that they have found and perfected ways and means of opposing, for a time, the forces that tend to the dissipation of energy. Life is a revolt against inertia. Those species that have died out are extinct, either because they came to the end of their resources, or because they became so inflexibly adapted to a certain kind of life that they were unable to meet the emergency of a change in the conditions that made this life possible. Efficiency in the ordinary means of living, rather than specialization for a particular way of living, appears to be the best guarantee of continued existence.

CHAPTER V

TERMITES

Iᴛ was the custom, not long ago, to teach the inexperienced that the will can achieve whatever ambition may desire. "Believe that you can, and you can, if only you work hard enough"; this was the subject of many a maxim very encouraging, no doubt, to the young adventurer, but just as likely to lead to a bench in Union Square as to a Fifth Avenue studio or a seat in the Stock Exchange.

Now it is the fashion to give us mental tests and vocational suggestions, and we are admonished that it is no use trying to be one thing if nature has made us for something else. This is sound advice; the only trouble is the difficulty of being able to detect at an early age the characters that are to distinguish a plumber from a doctor, a cook from an actress, or a financier from an entomologist. Of course, there really are differences between all classes of people from the time they are born, and a fine thing it would be if we could know in our youth just what each one of us is designed to become. In the present chapter we are to learn that certain insects appear to have achieved this very thing.

The termites are social insects; consequently in studying them, we shall be confronted with questions of conduct. Therefore, it will be well at the outset to look somewhat into the subject of morality; not, be assured, to learn any of its irksome precepts, but to discover its biological significance.

Right and wrong, some people think, are general abstractions that exist in the very nature of things. They are, on the contrary, specific attributes that are conditioned by circumstances. An act that is right is one in accord with the nature of the creature performing it; that which is wrong is a contrary act. Hence, what is right for one species of animal may be wrong for another, and the reverse.

The conduct of adult human individuals, according to human standards of right and wrong, we call morals; the similar conduct of other animals is a part of what biologists call *behavior*. But we unconsciously recognize something in common between morals and behavior when we speak of the acts of a child, which we call his behavior rather than his morals. Behavior, in other words, we regard as involving less of personal responsibility than morality. Hence we say that animals and children behave, but that adult human beings consciously do right or wrong. Yet, the two modes of action accomplish similar results: if the child behaves properly, his actions are right; if the adult has a properly developed moral sense, he too does the right thing, or at least he refrains from doing the wrong thing unless misguided by circumstances or by his reasoning.

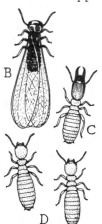

Fig. 75. A common species of termite of eastern North America inhabiting dead wood, *Reticulitermes flavipes*. A, B, winged forms. C, a soldier. D, workers.

Animals other than the human, it appears, generally do what is right from their standpoint; but their actions, we say, are instinctive. Some will insist that the terms "right" and "wrong" can have no application to them. Substitute then, if you please, the expression "appropriate or non-appropriate to the animal's way of living." And still, our morality will analyze into the same two elements; our acts are right or wrong according as they are appropriate or non-appropriate to *our* way of living.

The difference between human actions and those of other animals is not essentially in the acts themselves, but in the methods by which they are brought about. Animals are controlled by instincts, mostly; man is controlled by a conscious feeling that he should do this or that—"conscience," we call it—and his specific actions are the result of his reasoning or teaching as to what is right and what is wrong, excepting, of course, the acts of perverted individuals who lack either a functional conscience or a well-adjusted power of reason, or of individuals in whom the instincts of an earlier way of living are still strong. The general truth is clear, however, that in behavior, as in physiology, there is not just one way of arriving at a common result, and that nature may employ quite different means for determining and activating conduct in her creatures.

Since right and wrong, then, are not abstract properties, but are terms expressing fitness or non-fitness, judged according to circumstances, or an animal's way of living, it is evident that the quality of actions will differ much according to how a species lives. Particularly will there be a difference in the necessary behavior of species that live as individuals and of those that live as groups of individuals. In other words, that which may be right for an individualistic species may be wrong for a communal species; for, with the latter, the group replaces the individual, and relations are now established within the group, or pertaining to the group as a whole, that before applied to the individual, while relations that formerly existed between individuals become now relations between groups.

The majority of animals live as individuals, each wandering here and there, wherever its fancy leads or wherever the food supply attracts it, recognizing no ties or responsibilities to others of its species and contending with its fellows, often in deadly combat, for whatever advantage it can gain. A few animals are communistic or social in their mode of life; notably so are man and certain insects. The best-known examples of social insects are the ants and some of the bees and wasps. The termites, however, constitute another group of social insects of no less interest than the ants and bees, but whose habits have not been so long observed.

Fig. 76. Termite work in a piece of wood. Tunnels following the grain are made by species of *Reticulitermes*, the common underground termites of the eastern United States

More familiarly to some people, termites are known as "white ants." But since they are not ants, nor always white or even pale in color, we should discard this misleading and unjustifiable appellation and learn to know the termites by the name under which they are universally known to entomologists.

If you split open an old board that has been lying almost anywhere on the ground for some time, or if, when out in the woods, you cut into a dead stump or a log, you are more than likely to find it tunneled all through with small tubular galleries running with the grain of the wood, but everywhere connected crosswise by small openings or short passages. Within the exposed galleries there will be seen numerous small, pale, wingless insects running here and there in an effort to conceal themselves. These insects are termites. They are the miners or the descendants of miners that have excavated the tunnels in which they live. Not all of the galleries in the nest are open runways, many of them being packed solidly with small pellets of refuse.

If the termites confined themselves to useless wood, they would be known only as interesting insects; but since they often extend their operations into fence posts, telegraph poles, the woodwork of houses, and even into furniture, they have placed themselves among the destructive insects and have acquired an important place in the pages of economic entomology. Stored papers, books, cloth, and leather are not exempt from their attack. In the United States it not infrequently happens that the flooring or other wooden parts of buildings must be replaced, owing to the unsuspected work of termites; and piled lumber is especially liable to invasion by these insidious insects. But in tropical countries the termites are far more numerous than in temperate regions, and are vastly more destructive than they are with us. Their seclusive habits make the termites a particularly vexatious pest, because they have usually accomplished an irreparable amount of damage before their presence is known or suspected. The economic entomologist studying termites gives most of his attention, therefore, to devising methods of preventing the access of the insects to all wooden structures that they might destroy.

The work of termites and the ways and means that have been contrived to prevent their ravages have been described in many agricultural publications, and the reader whose tastes are purely practical is referred to

the latter for information. Here we will look more closely
into the lives of the termites themselves to see what
lessons we may learn from these creatures that have
adopted something of our own way of living.

When a termite nest is broken open, it does not appear
that there is much of an organization among the insects

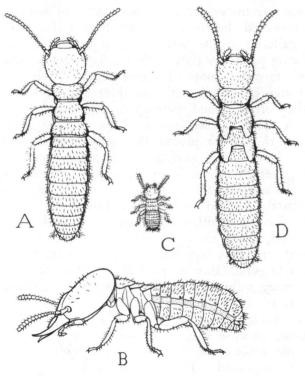

Fig. 77. *Reticulitermes flavipes* (much enlarged)
A, a mature worker. B, a mature soldier. C, a young termite.
D, an immature winged form

hurrying to take refuge in the recesses of the galleries,
but neither when a bomb strikes one of our own dwellings
is there probably much evidence of order within. The
most casual observation of the termites, however, will

show something of interest concerning them. In the first place, it is to be seen that not all the members of the colony are alike. Some, usually the greater number, are small, ordinary, soft-bodied, wingless insects with rounded heads and inconspicuous jaws (Figs. 75 D, 77 A). Others, less numerous, have bodies like the first, and are also wingless, but their heads are relatively of enormous size and support a pair of large, strong jaws projecting out in front (Figs. 75 C, 77 B). The individuals of the latter kind are known as *soldiers*, and the name is not entirely fanciful, since fighting is not necessarily the everyday occupation of one in military service. The others, the small-headed individuals, are called *workers*, and they earn their title literally, for, even with their small jaws, they do most of the work of excavating the tunnels, and they perform whatever other labors are to be done within the nest.

Both the workers and the soldiers are males and females, but so far as reproductive powers go, they may be called "neuters," since their reproductive organs never mature and they take no part in the replenishment of the colony. In most species of termites the workers and the soldiers are blind, having no eyes or but rudiments of eyes. In a few of the more primitive termite genera, workers are absent, and in the higher genera they may be of two types of structure. The large jaws of the soldiers (Fig. 78 A) are weapons of defense in some species, and the soldiers are said to present themselves at any break in the walls of the nest ready to defend the colony against invasion. In some species, the soldiers have a long tubular horn projecting forward from the face (Fig. 78 B), through which opens the duct of a gland that emits a sticky, semiliquid substance. This glue is discharged upon an attacking enemy, who is generally an ant, and so thoroughly gums him up that he is rendered helpless—a means of combat yet to be adopted in human warfare. The facial gland is developed to such efficiency as a weapon in many species of one termite family that the soldiers of these species have no need of jaws, and their mandibles have become rudimentary. In all cases, the military specialization of the soldiers has rendered them incapable of feeding themselves, and they must depend on the workers for food.

In addition to the soldiers and the workers, there would probably be seen within the termite nest, at cer-

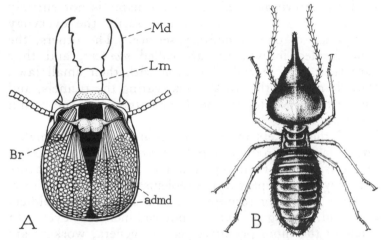

Fig. 78. Two forms of defensive organs of termite soldiers

A, head of soldier of *Termopsis*, showing the highly developed mandibles (*Md*), and the great muscles within the head (*admd*) that close them. B, a soldier of *Nasutitermes* (from Banks and Snyder); the head has small jaws but is provided with a long snoutlike horn through which is ejected a gummy liquid used for defense

tain seasons of the year, many individuals (Fig. 77 D) that have small wing rudiments on their thoracic segments. As the season advances, the wing pads of these individuals increase in length, until at last they become long, gauzy, fully-developed wings extending much beyond the tip of the body (Figs. 75 A, B, 79). The color of the body also becomes darker, and finally blackish when the insects are mature. Then, on some particular day, the whole winged brood issues from the nest in a great swarm. Since insects are normally winged creatures, it is evident that these flying termites represent the perfect forms of the termite colony—they are, in fact, the sexually mature *males* and *females*.

The several forms of individuals in the termite community are known as *castes*.

An intensive search through the galleries of a termite nest might reveal, besides workers, soldiers, and the members of the winged brood in various stages of development, a few individuals of still different kinds. These have heads like the winged forms, but rather larger bodies; some have short wing rudiments (Fig. 80), others have none; and finally there are two individuals, a male and a female, bearing wing stubs from which, evidently, fully-formed wings have been broken off. The male of this last pair is just an ordinary-looking, though dark-bodied termite (Fig. 82 A); but the female is distinguished from all the other members of the colony by the great size of her abdomen (B).

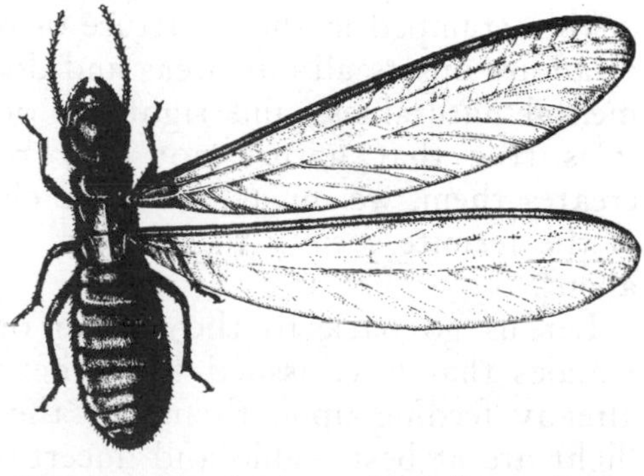

Through the investigations of entomologists it is known that the short-winged and wingless individu-

FIG. 79. Adult winged caste of *Reticulitermes tibialis*, wings shown on one side of the body only. (From Banks and Snyder)

als of this group comprise both males and females that are potentially capable of reproduction, but that in general all the eggs of the colony are actually produced by the large-bodied female, whose consort is the male that has lost his wings. In other words, this fertile female corresponds with the "queen" in a hive of bees; but, unlike the queen bee, the queen termite allows the "king" termite to live with her throughout her life in the community.

It appears, then, that the termite community is a complex society of castes, for we must now add to the worker and soldier castes the two castes of potentially reproductive individuals, and the "royal" or actual producing caste, consisting of the king and the queen. We are thus introduced to a social state quite different from anything known in our own civilization, for, though we may have castes, the distinctions between them are largely matters of polite concession by the less aspiring members of the community. We theoretically claim that we are all born equal. Though we know that this is but a gratifying illusion, our inequalities at least do not go by recognized caste. A termite, however, is literally born into his place in society and eventually has his caste insignia indelibly stamped in the structure of his body. This state of affairs upsets all our ideas and doctrines of the fundamental naturalness and rightness of democracy; and, if it is true that nature not only recognizes castes but creates them, we must look more closely into the affairs of the termite society to see how such things may come about.

Let us go back to the swarm of winged males and females that have issued from the nest. The birds are already feeding upon them, for the termites' powers of flight are at best feeble and uncertain. The winds have scattered them, and in a short time the fluttering horde will be dispersed and probably most of its members will be destroyed one way or another. The object of the swarming, however, is the distribution of the insects, and, if a few survive, that is all that will be necessary for the continuance of the race. When the fluttering insects alight they no longer have need of their wings, and by brushing against objects, or by twisting the body until the tip of the abdomen comes against the wing bases, the encumbering organs are broken off. It may be observed that there is a suture across the base of each wing just to make the breaking easy.

The now wingless termites, being young males and females just come to maturity, naturally pair off; but

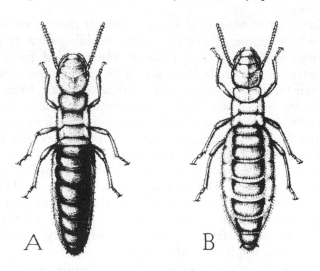

FIG. 80. The second form, or short-winged reproductive caste, of *Reticulitermes tibialis*. (From Banks and Snyder)
A, male. B, female

not for a companionate marriage, which, it must be confessed, is the popular form of matrimony with most insects. The termites take the vows of lifetime fidelity, or "till death do us part," for with the female termite intensive domesticity and maternity are the ruling passions. To find a home site and there found a colony is her consuming ambition, and, whether the male likes it or not, he must accept her conditions. The female, therefore, searches out a hole or a crevice in a dead tree or a decayed stump, or crawls under a piece of wood lying on the ground, and the male follows. If the site proves suitable, the female begins digging into the wood or into the ground beneath it, using her jaws as excavating tools, perhaps helped a little by the male, and soon a shaft is sunk at the end of which a cavity is hollowed out of sufficient size to accommodate the pair and to serve the purposes of a nest where true matrimony may begin.

Naturally it would be a very difficult matter to follow the whole course of events in the building of a termite community from one of these newly married pairs, for the termites live in absolute seclusion and any disturbance of their nests breaks up the routine of their lives and frustrates the efforts of the investigator. Many phases, however, of the life and habits of our common eastern United States termites, particularly of species belonging to the genus *Reticulitermes*, have been discovered and recorded in numerous papers by Dr. T. E. Snyder of the U. S. Bureau of Entomology, and, thanks to Doctor Snyder's work, we are able to give the following account of the life of these termites and the history of the development of a fairly complex community from the progeny of a single pair of insects.

The young married couple live amicably together in conjugal relations within their narrow cell. The male, perhaps, was forced to eject a would-be rival or two, but eventually the mouth of the tunnel is permanently sealed, and from now on the lives of this pair will be completely shut in from the outside world. In due time, a month or six weeks after the mating, the female lays her first eggs, six or a dozen of them, deposited in a mass on the floor of the chamber. About ten days thereafter the eggs hatch, and the new home becomes enlivened with a brood of little termites.

The young termites, though active and able to run about, are not capable of feeding themselves, and the parents are now confronted with the task of keeping a dozen growing appetites appeased. The feeding formula of the termite nursery calls for predigested wood pulp; but fortunately this does not have to be supplied from outside—the walls of the house furnish an abundance of raw material and the digesting is done in the stomachs of the parents. The pulp needs then only to be regurgitated and handed to the infants. This feature in the termite economy has a double convenience, for not only are the young inexpensively fed, but the gathering of the food automatically enlarges the home to accommodate the increasing need for space of the growing family.

That insects should gnaw tunnels through dead wood is not surprising; but that they should be able to subsist on sawdust is a truly remarkable thing and a dietetic feat that few other animals could perform. Dry wood consists mostly of a substance called cellulose, which, while it is related to the starches and sugars, is a carbohydrate that is entirely indigestible to ordinary animals, though eaten in abundance as a part of all vegetable food. The termites, however, are unusually gifted, not with a special digestive enzyme, but with minute, one-celled, cellulose-digesting protozoan parasites that live in their alimentary canals. It is through the agency of their intestinal inhabitants, then, that the termites are able to live on a diet of dead wood. The young termites receive some of the organisms with the food given them by their parents and are soon able to be wood eaters themselves. Not all termites, however, are known to possess these intestinal protozoa, and, as we shall see, many of them feed on other things than wood.

The termite brood thrives upon its wood-pulp diet, and by December following the spring in which the young were hatched, the members of the new generation begin to attain maturity after having progressed through a series of moltings, as does any other growing insect. But observe, the individuals of this generation, instead of developing into replicas of their parents, have taken on the form of workers and soldiers! However, one should never express surprise when dealing with insects; and for the present we must accept the strange development of the young termites as a matter of fact, and pass on.

During the middle of winter things remain thus in the new family colony. The members of termite species that live in the ground, or that pass from wood into the ground, probably have tunneled deep into the earth for protection from the cold. But in February, the mother termite, now the queen of the brood, responds again to the urge of maternity with some more eggs, probably with a greater number this time than on the first occasion. A month later, or during March, the termitary is once more enlivened with young termites. The king and the queen are now, however, relieved of the routine of nursery duties by the workers of the first brood. The latter take over the feeding and care of their new brothers and sisters, and also do all the excavation work involved in the enlarging of the home.

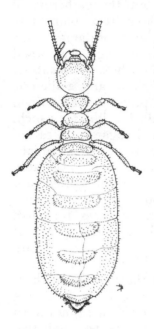

FIG. 81. A queen of the third form, or wingless reproductive caste, of *Reticulitermes flavipes*. (From Banks and Snyder)

In the spring the termites ascend to the surface of the ground beneath a board or log, or at the base of a stump, and reoccupy their former habitation. As the galleries are extended, the family moves along, slowly migrating thus to uneaten parts of the wood and leaving the old tunnels behind them mostly packed with excreted wood-pulp and earth.

When June comes again, the young family may consist of several dozen individuals; but all, except the king and queen, are soldiers and workers, the latter much out-numbering the former. During the second year, the queen lays a still greater number of eggs and probably produces them at more frequent intervals. With the increase in the activity of her ovaries, her abdomen enlarges and she takes on a matronly appearance, attaining a length fully twice that of her virgin figure and a girth in proportion. The king, however, remains faithful to his spouse; and he, too, may fatten up a little, sufficiently to give him some distinction amongst his multiplying subjects. The termite king is truly a king, in the modern way, for he has renounced all authority and responsibility and leads a care-free life, observing only the decorums of polite society and adhering to the traditions of a gentleman; but he also achieves the highest distinction of democracy, for he is literally the father of his country.

Another year rolls by, bringing more eggs, more workers, more soldiers. And now, perhaps, other forms appear in the maturing broods. These are marked at a certain stage of their development by the possession of short wing stubs or pads on the back of the normally wing-bearing segments. With succeeding molts the wing pads become larger and larger, until they finally develop, in most of these individuals, into long wings like those of the king and the queen when they first flew out from the parent colony. At last, then, the new family is to have its first swarm; and when the fully-winged members are all ready for the event and the proper kind of day arrives, the workers open a few exits from the galleries, and the winged ones are off. We already know their history, for they will only do what their parents did before them and what their ancestors have done for millions of generations. Let us go back to the galleries.

A few of the individuals that developed winged pads are fated to disappointment, for their wings never grow to a functional size and they are thereby prevented from joining the swarm. Their reproductive organs and their instincts, however, attain maturity, and these short-winged individuals, therefore, become males and females capable of procreation. They differ from the fully-winged sexual forms in a few respects other than the length of the wings, and they constitute a true caste of the termite community, that of the *short-winged males and females* (Fig. 80). The members of this caste mature along with the others, and, Doctor Snyder tells us, many of them, regardless of their handicap, actually leave the nest at the time the long-winged caste is swarming; as if in them, too, the instinct for flight is felt, though the organs for accomplishing it are unable to play their part. Just what becomes of these unfortunates is a mystery, for Doctor Snyder says that after the swarming none of them is to be found in the nest. It may be that some of them pair and found new colonies after the manner of the winged forms, but the facts concerning their history are not known. It is at least true that colonies are sometimes found which have no true royal pair, but in which the propagating individuals are members of this short-winged reproductive caste.

Finally, there are also found in the termite colonies certain wingless individuals that otherwise resemble the winged forms, and which, as the latter, are functionally capable of reproduction when mature. These individuals constitute a third reproductive caste—the *wingless males and females*. Little is known of the members of this caste, but it is surmised that they may leave the nests by subterranean passages and found new colonies of their own.

Just how long the primary queen of a colony can keep on laying eggs is not known, but in the course of years she normally comes to the end of her resources, and before that time she may be injured or killed through some accident. Her death in any case, however, does not mean the end of the colony, for the king may provide for the continuance of his race, and at the same time console himself in his bereavement, by the adoption of a whole harem of young short-winged females. But if he too should be lost, then the workers give the succession to one or more pairs of the second- or third-caste reproductive forms, to whom they grant the royal prerogatives. The progeny of any of the fertile castes will include the caste of the parents and all castes below them. In other words, only winged forms can produce the whole series of castes; short-winged parents can not produce long-winged offspring; and wingless parents can not produce winged

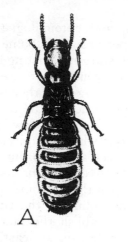
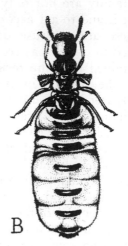

FIG. 82. The usual king (A) and queen (B), or winged reproductive caste after having lost the wings (fig. 79), of *Reticulitermes flavipes*. (From Banks and Snyder)

offspring of any form; but both short-winged and wingless parents can produce soldiers and workers. It appears, therefore, that each imperfect fertile insect lacks something in its constitution that is necessary for the production of a complete termite individual.

The production of constitutionally different castes from the eggs of a single pair of parents would be a highly disconcerting event if it happened anywhere else than in a termite colony, where it is the regular thing. But the fact of its being regular with termites makes it none the less disconcerting to entomologists, for it seems to defy the very laws of heredity.

There can be no doubt of the utility of a caste system where the members of each caste know their places and their duties, and where nobody ever thinks of starting a social revolution. But we should like to know how such a system was ever established, and how individuals of a family are not only born different but are made to admit it and to act accordingly.

These are abstruse questions, and entomologists are divided in opinion as to the proper answers. Some have maintained that the termite castes are not distinguished when the various individuals are young, but are produced later by differences in the feeding—in other words, it is claimed the castes are made to order by the termites themselves. One particular objection to this view is that no one has succeeded in finding out what the miraculous pabulum may be, and no one has been able to bring about a structural change in any termite by controlling its diet. On the other hand, it has been shown that in some species there are actual differences in the young at the time of hatching, and such observations establish the fact that insects from eggs laid by one female *can*, at least, give rise to offspring of two or more forms, beside those of sex, and that potential differences are determined in the eggs. It is most probable that in these forms no structural differences could be discovered at an early embryonic period, and hence it may be that, where differences are not perceptible at the time of hatching, the period of differentiation has only been delayed to a later stage of growth. It is possible that a solution to the problem of the termite castes will be found when a study of the eggs themselves has been made.

We may conclude, therefore, that the structural differences between the termite castes are probably innate, and that they arise from differences in the constitutional elements of the germ cells that direct the subsequent development of the embryos in the eggs and of the young after hatching.

Still, however, there remain questions as to the nature of the force that controls termite behavior. Why do the termites remain together in a community instead of scattering, each to live its own life as do most other insects? Why do the workers accept their lot and perform all the menial duties assigned to them? Why do the soldiers expose themselves to danger as defenders of the nests? Structure can account for the things it is impossible for an animal to do, but it can not explain *positive* behavior where seemingly the animal makes a choice between many lines of possible action open to it.

In the community of the cells that make up the body of an animal, as we learned in Chapter IV, organization and control are brought about either through the nerves, which transmit an activating or inhibiting force to each cell from a central controlling station, or through chemical substances thrown into the blood. In the insect community, however, there is nothing corresponding to either of these regulating influences; nor is there a law-making individual or group of individuals as in human societies, nor a police force to execute the orders if any were issued. It would seem that there must be some inscrutable power that maintains law and order in the termite galleries. Are we, then, to admit that there is a "spirit of the nest," an "*âme collective*," as Maeterlinck would have us believe—some pervading force that unites the individuals and guides the destinies of the colony as a whole? No, scientists can not accept any such idea as that, because it assumes that nature's resources are no greater than those of man's imagination. Nature is always natural, and her ways and means of accomplishing anything, *when once discovered*, never invoke things that the human mind can not grasp, except in their ultimate analysis into first principles. Those who have faith in the consistency of nature endeavor to push a little farther into the great unknown knowable.

There are a few things known about the termites that help to explain some of the apparent mysteries concerning them. For example, the members of a colony are forever licking or nibbling at one another; the workers appear to be always cleaning the queen, and they are assiduous in stroking the young. These labial attentions, or lip affections, moreover, are not unrewarded, for it appears that each member of the colony exudes some substance through its skin that is highly agreeable to the other members. Furthermore, the termites all feed one another with food material ejected from the alimentary canal, sometimes from one end, sometimes from the other. Each individual, therefore, is a triple source of nourishment to his fellows—he has to offer exudates from the skin, crop food from the mouth, and intestinal food from the anus—and this mutual exchange of food appears to form the basis for much of the attachment that exists among the members of the colony. It accounts for the maternal affections, the care of the queen and the young by the workers, the brotherly love between the workers and the soldiers. The golden rule of the termite colony is "feed others as you would be fed by them."

The termites, therefore, are social creatures because, for physical reasons, no individual could live and be happy away from his fellows. The same might be said of us, though, of course, we like to believe that our social instincts have not a purely physical basis. Be that as it may, we must recognize that any kind of social tie is but one of various possible means by which the benefits of community life are insured to the members of the community.

The custom of food exchange in the termite colonies can not be held to account by any means for all the things that termites do. Where other explanations fail, we have always to fall back on "instinct." A true instinct is a response bred in the nervous system; and the behavior of termites, as of all other insects, is largely brought

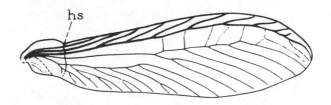

hs

FIG. 83. A fore wing of a termite, *Kalotermes approximatus*, showing the humeral suture (*hs*) where the wing breaks off when it is discarded

about by automatic reflexes that come into action when external and internal conditions are right for their production. The physical qualities of the nervous system that make certain reactions automatic and inevitable are inherited; they are transmitted from parent to off-spring, and bring about all those features of the animal's behavior that are repeated from generation to generation and which are not to be attributed to the individual's response to environmental changes.

The termites have an ancient lineage, for though no traces of their family have been found in the earlier records, there can be no doubt that the ancestors of the termites were closely related to those of the roaches; and the roach family, as we have seen in Chapter III, may be reckoned among the very oldest of winged insects. In human society it means a great deal to belong to an "old family," at least to the members of that family; but in biology generally it is the newer forms, the upstarts of more recent times, that attain the highest degree of organization; and most of the social insects—the ants, the bees, and the wasps—belong to families of comparatively recent origin. It is refreshing, therefore, to find the belief in aristocracy vindicated by the ancient and honorable line of descent represented by the roaches and flowering in the termites.

One particular piece of evidence of the roach ancestry of the termites is furnished by the wings. With most termites the wings (Fig. 83) are not well developed, and

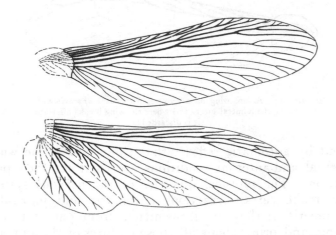

FIG. 84. Wings of *Mastotermes*, the hind wing with a basal expansion similar to that of the hind wing of a roach (fig. 53), suggesting a relationship between termites and roaches

their muscles are partly degenerate. In some forms, however, the wings (Fig. 84) are distinctly of the roach type of structure (Fig. 53), and these forms are undoubtedly more closely representative of the ancestral termites than are the species with the usual termite wing structure.

Our termites and those of other temperate regions constitute the mere fringes of termite civilization. The termites are particularly insects of warm climates, and it is in the tropics that they find their most congenial environment and attain the full expression of their possibilities.

In the tropics the characteristic termites are not those that inhabit dead wood, but species that construct definite and permanent nests, some placed beneath the ground, others reared above the surface, and still others built against the trunks or branches of trees. Different species employ different building materials in the construction of their nests. Some use particles of earth, sand grains, or clay; others use earth mixed with saliva; still others make use of the partly digested wood pulp ejected from their bodies; and some use mixed materials. Certain kinds of tropical termites, moreover, have foraging habits.

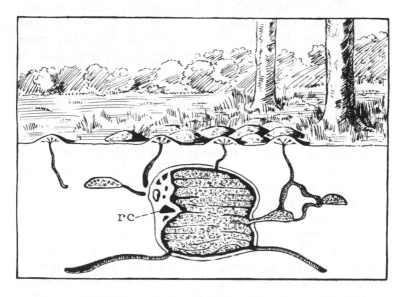

Fig. 85. Vertical section of an underground nest of an African termite, *Termes badius*. (From Hegh, after Fuller)

The large central chamber is the principal "fungus garden"; in the wall at the left is the royal chamber (*rc*); tunnels lead from the main part of the nest to smaller chambers containing fungus, and to the small mounds at the surface

Great armies of workers of these species leave the nests, even in broad daylight, and march in wide columns guarded by the soldiers to the foraging grounds, where they gather bits of leaves, dead stems, or lichens, and return laden with provender for home consumption.

The underground nests (Fig. 85) consist chiefly of a

cavity in the earth, perhaps two by three feet in diameter and a foot beneath the surface, walled with a thick cement lining; but from this chamber there may extend tunnels upward to the surface, or horizontally to other smaller chambers located at a distance from the central one. The termites that live in these nests subsist principally upon home-grown food, and it is in the great vaulted central chamber that they raise the staple article of their diet. The cavity is filled almost entirely with a porous, spongy mass of living fungus. The fungi as we ordinarily see them are the toadstools and mushrooms, but these fungus forms are merely the fruiting bodies sent up from a part of the plant concealed beneath the ground or in the dead wood; and this hidden part has the form of a network of fine, branching threads, called a *mycelium*. The mycelium lives on decaying wood, and it is the mycelial part of the fungus that the termites cultivate. They feed on small spore-bearing stalks that sprout from the threads of the mycelium. The substratum of the termite fungus beds is generally made of pellets of partly digested wood pulp.

The nests that termites erect above the ground include the most remarkable architectural structures produced by insects. They are found in South America, Australia, and particularly in Africa. In size they vary from mere turrets a few inches high to great edifices six, twelve, or even twenty feet in altitude. Some are simple mounds (Fig. 86 A), or mere hillocks; others have the form of towers, obelisks, and pyramids (B); still others look like fantastic cathedrals with buttressed walls and tapering spires (Fig. 87); while lastly, the strangest of all resemble huge toadstools with thick cylindrical stalks and broad-brimmed caps (Fig. 86 C). Many of the termites that build mound nests are also fungus-growing species, and one chamber or several chambers in the nest are given over to the fungus culture.

Termite nests built in trees are usually outlying retreats

of colonies that live in the ground, for such nests (Fig. 86 D) are connected with an underground nest by covered runways extending down the trunk of the tree.

The queens of nearly all the termites that live in permanent nests attain an enormous size by the growth of the abdomen, the body becoming thus so huge that the royal

FIG. 86.   Four common types of above-ground nests made by tropical termites A, type of small mound nest, varying from a few inches to several feet in height. B, type of a large tower or steeple nest, reaching a height of 9 or 10 feet.   C, a mushroom-shaped nest, made by certain African termites, from 3 to 16 inches high.   D, a tree nest, showing the covered runway going down to the ground

female is rendered completely helpless, and must be attended in all her wants by the workers.   With such species the queen is housed in a special royal chamber which she never leaves.   Her body becomes practically a great bag in which the eggs are produced, and so great is the fertility of one of these queens that the ripened eggs continually issue from her body.   It has been estimated that in one such species the queen lays four thousand eggs a day, and that in another species her daily output may be thirty thousand.   Ten million eggs a year is pos-

Fig. 87. Type of pinnacled nest made by species of African termites, sometimes reaching a height of twenty feet or more

sibly a world record in ovulation. The royal chamber is usually placed near the fungus gardens, and as fast as the eggs are delivered by the queen the attendant workers carry them off to the garden and distribute them over the fungus beds, where the young on hatching can feed and grow without further attention.

From a study of the termites we may draw a few lessons for ourselves. In the first place, we see that the social form of life is only one of the ways of living; but that, wherever it is adopted, it involves an interdependence of individuals upon one another. The social or community way of living is best promoted by a division of labor among groups of individuals, allowing each to specialize and thereby to attain proficiency in his particular kind of work. The means by which the termites have achieved the benefits of social life are not the same as those adopted by the ants or social bees, and they have little in common with the principles of our own social organization. All of which goes to show that in the social world, as in the physical world, the end alone justifies the means, so far as nature is concerned. Justice to the individual is a human concept; we strive to equalize the benefits and hardships of the social form of life, and in so far as we achieve this aim our civilization differs from that of the insects.

CHAPTER VI

PLANT LICE

"Plant lice! Ugh," you say, "who wants to read about those nasty things! All I want to know is how to get rid of them." Yes, but the very fact that those soft green bugs that cover your roses, your nasturtiums, your cabbages, and your fruit trees at certain seasons reappear so persistently, after you think you have exterminated them, shows that they possess some hidden source of power; and the secrets of a resourceful enemy are at least worth knowing—besides, they may be interesting.

Really, however, insects are not our enemies; they are only living their appointed lives, and it just happens that we want to eat some of the same plants that they and their ancestors have always fed on. Our trouble with the insects is just that same old economic conflict that has bred the majority of wars; and, in the case between us and the insects, it is we who are the aggressors and the enemies of the insects. We are the newcomers on the earth, but we fume around because we find it already occupied by a host of other creatures, and we ask what right have they to be here to interfere with us! Insects existed millions of years before we attained the human form and aspirations, and they have a perfectly legitimate right to everything they feed on. Of course, it must be admitted, they do not respect the rights of private property; and therein lies their hard luck, and ours.

The plant lice are well known to anyone who has a garden, a greenhouse, an orchard, or a field of grain. Some call them "green bugs"; entomologists usually call

them *aphids*. A single plant louse is an *aphis*, or an *aphid;*
more than one are usually called *aphides*, or *aphids*.

The distinguishing feature of the plant lice, or aphids,
as we shall by preference call them, is their manner of feeding. All the insects described in the preceding chapters
eat in the usual fashion of biting off pieces of their food,
chewing them, and swallowing the masticated bits. The

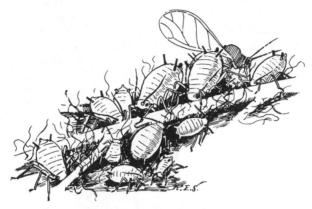

FIG. 88.   Group of green apple aphids feeding along a rib on
under surface of an apple leaf

aphids are sucking insects; they feed on the juices of the
plants they inhabit. Instead of jaws, they have a piercing
and sucking beak (Fig. 89), consisting of an outer sheath
inclosing four slender, sharp-pointed bristles which can be
thrust deep into the tissues of a leaf or stem (Fig. 89 B).
Between the bristles of the innermost pair (Fig. 90, *Mx*)
are two canals. Through one canal, the lower one (*b*), a
liquid secretion from glands of the head is injected into the
plant, perhaps breaking down its tissues; through the
other (*a*) the plant sap and probably some of the protoplasmic contents of the plant cells are drawn up into the
mouth. A sucking apparatus like that of the aphids is
possessed by all insects related to the aphids, comprising
the order *Hemiptera*, and will be more fully described

[153]

in the next chapter, which treats of the cicada, a large cousin of the aphids.

When we observe, now, that different insects feed in two quite different ways, some by means of the biting type of mouth parts, and others by means of the sucking type, it becomes evident that we must know which kind of insect we are dealing with in the case of pests we may be trying to control. A biting and chewing insect can be killed by the mere expedient of putting poison on the outside of its food, if it does not become aware of the poison and desist from eating it; but this method would not work with the piercing and sucking insects, which extract their food from beneath the surface of the plants on which they feed. Sucking insects are, therefore, to be destroyed by means of sprays or dusts that will kill them by contact with their bodies. Aphids are usually attacked with irritant sprays, and in general it is not a difficult matter to rid infested plants of them, though in most cases the spraying must be repeated through the season.

When any species of aphis becomes well established on a plant, the infested leaves (Fig. 88) may be almost as crowded as an East Side street on a hot summer afternoon. But there is

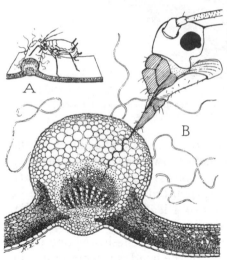

FIG. 89. The way an aphis feeds on the juices of a plant

A, an aphis with its beak thrust into a rib of a leaf. B, section through the midrib of a young apple leaf, showing the mouth bristles from the beak of an aphis penetrating between the cells of the leaf tissue to the vascular bundles, while the sheath of the beak is retracted by folding back beneath the head

PLATE 2

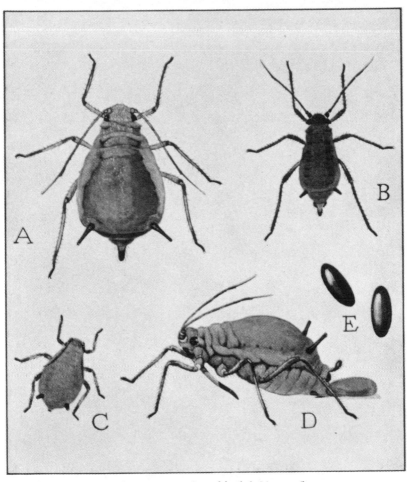

The green apple aphis (*Aphis pomi*)
A, adult sexual female; B, adult male; C, young female; D, female laying an egg; E, eggs, which turn from green to black after they are laid.
(Enlarged about 20 times)

no bustle, no commotion, for each insect has its sucking bristles buried in the leaf, and its pump is busy keeping the stomach supplied with liquid food. The aphis crowds are mere herds, not communities or social groups as in the case of the termites, ants, or bees.

Wherever there are aphids there are ants, and in contrast to the aphids, the ants are always rushing about all over the place as if they were looking for something and each wanted to be the first to find it. Suddenly one spies a droplet of some clear liquid lying on the leaf and gobbles it up, swallowing it so quickly that the spherule seems to vanish by magic, and then the ant is off again in the same excited manner. The explanation of the presence and the actions of the ants among the aphids is this: the sap of the plants furnishes an unbalanced diet, the sugar content being far too great in proportion to the protein. Consequently the aphids eject from their bodies drops of sweet liquid, and it is this liquid, called "honey dew," that the ants search out so eagerly. Some of the ants induce the aphids to give up the honey dew by stroking the bodies of the latter. The glistening coat often seen on the leaves of city shade trees and the shiny liquid that bespatters the sidewalks beneath is honey dew discharged from innumerable aphids infesting the under surfaces of the leaves.

In studying the termites, we learned that it is possible for a single pair of insects to produce regularly several kinds of offspring differing in other ways than those of sex. In the aphids, a somewhat similar thing occurs in that each species may be represented by a number of forms; but with the aphids these different forms constitute successive generations. If events took place in a human family as they do in an aphid family, children born of normal parents would grow up to be quite different from either their father or their mother; the children of these children would again be different from their parents and also from their grandparents, and when mature they perhaps would migrate to some other part of the country; here they would have children of their own, and the new fourth generation would be unlike any of the three preceding; this generation would then produce another, again different; and the latter would return to the home town of their grandparents and great-grandparents, and here bring forth children that would grow up in the likeness of their great-great-great-grandparents! This seems like a fantastic tale of fiction, too preposterous to be taken seriously, but it is a commonplace fact among the aphids, and the actual genealogy may be even more complicated than that above outlined. Moreover, the story is not yet complete, for it must be added that all the generations of the aphids, except one in each series, are composed entirely of females capable in themselves of reproduction. In warm climates, it appears, the female succession may be uninterrupted.

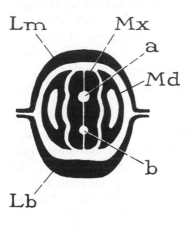

FIG. 90. Cross-section through the base of the beak of an aphis. (From Davidson)

The outer sheath of the beak is the labium (*Lb*), covered basally by the labrum (*Lm*). The four inclosed bristles are the mandibles (*Md*) and the maxillae (*Mx*), the latter containing between them a food canal (*a*) and a salivary canal (*b*). Only the inner walls of the labrum and labium are shown in the section

How insects do upset our generalizations and our peace of mind! We have heard of feminist reformers who would abolish men. With patient scorn we have listened to their predictions of a millenium where males will be unknown and unneeded—and here the insects show us not only that the thing is possible but that it is practicable, at least for a certain length of time, and that the time can be indefinitely extended under favorable conditions.

Since special cases are always more convincing than general statements, let us follow the seasonal history of some particular aphids, taking as examples the species that commonly infest the apple.

Let the time be a day in the early part of March. Probably a raw, gusty wind is blowing from the northwest, and only the silver maples with their dark purplish clusters of frowzy flowers already open give any suggestion of the approach of spring. Find an old apple tree somewhere that has not been sprayed, the kind of tree an entomologist always likes to have around, since it is sure to be full of insects. Look closely at the ends of some of the twigs and you will probably find a number of little shiny black things stuck close to the bark, especially about the bases of the buds, or tucked under the projecting edges of scars and tiny crevices (Fig. 91). Each little speck is oval and about one thirty-sixth of an inch in length.

To the touch the objects are firm, but elastic, and if you puncture one a pulpy liquid issues from it; or so it appears, at least, to the naked eye— a microscope would show that in this liquid there is

FIG. 91. Aphis eggs on apple twigs in March; an enlarged egg below

organization. In short, the tiny capsule contains a young aphid, because it is an aphid egg. The egg was deposited on the twig last fall by a female aphis, and its living contents have remained alive since then, though fully exposed to the inclemencies of winter.

Immediately after being laid in the fall, the germ nucleus of the aphis egg begins development, and soon forms a band of tissue lying lengthwise on the under surface of the yolk. Then this scarcely-formed embryo undergoes a curious process of revolution in the egg, turning on a crosswise axis head foremost into the yolk and finally stretching out within the latter with the back down and the head toward the original rear end of the egg. Thus it remains through the winter. In March it again becomes active, reverses itself to its first position, and now completes its development.

The date of hatching of the apple aphis eggs depends much upon the weather and will vary, therefore, according to the season, the elevation, and the latitude; but in latitudes from that of Washington north, it is some time in April, usually from the first to the third week of the month. The eggs of most insects resemble seeds in their capacity for lying inert until proper conditions of warmth and moisture bring forth the creature biding its time within. The eggs of one of the apple aphids, however, are killed by premature warm weather, or if artificially warmed too long before the normal time of hatching. In general, the final development of the aphis embryos keeps pace with the development of the apple buds, since both are controlled by the same weather conditions, and this coordination usually insures the young aphids against starvation; but the eggs commonly hatch a little in advance of the opening of the buds, and a subsequent spell of cold

FIG. 92. Eggs of the green apple aphis with outer coverings split before hatching; below, an egg removed from its covering

[158]

weather may give the young lice a long wait for their first meal.

The approaching time of hatching is signaled in most

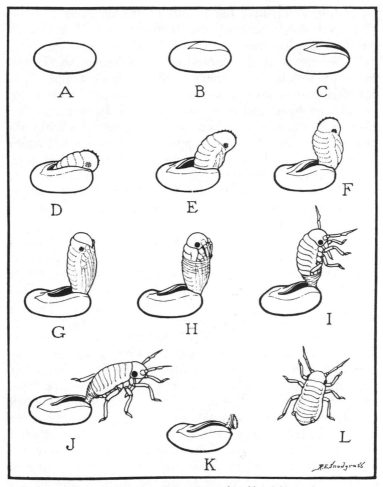

FIG. 93. Hatching of the green apple aphis, *Aphis pomi*
A, the egg. B, an egg with the outer coat split. C, the same egg with the inner shell split at one end. D–F, three successive stages in the emergence of the young insect. G–J, shedding the hatching membrane. K, the empty eggshell. L, the young aphid

[159]

INSECTS

cases by the splitting of an outer sheath of the egg (Fig.
92), exposing the glistening, black, true shell of the egg
within. Then, from one to several days later, the shell
itself shows a cleft within the rupture of the outer coat,
extending along half the length of the exposed egg surface and down around the forward end (Fig. 93 C).
From this split emerges the soft head of the young aphis
(D), bearing a hard, toothed crest, evidently the instrument by which the leathery shell was broken open, and
for this reason known as the "egg burster." Once exposed, the head continues to swell out farther and farther as if the creature had been compressed within the
egg. Soon the shoulders appear, and now the young
aphis begins squirming, bending, inflating its fore parts
and contracting its rear parts, until it works its body
mostly out of the egg (E, F) and stands finally upright
on the tip of its abdomen which is still held in the cleft
of the shell (G).

The young aphis at this stage, however, like the young
roach, is still inclosed in a thin, tight-fitting, membranous
bag having no pouches for the legs or other members,
which are all cramped within it. The closely swathed head
swells and contracts, especially the facial part, and suddenly the top of the bag splits close to the right side of
the egg burster (Fig. 93 H). The cleft pulls down over
the head, enlarges to a circle, slides along over the shoulders, and then slips down the body. As the tightly stretched
membrane rapidly contracts, the appendages are freed
and spring out from the body (I). The shrunken pellicle
is reduced at last to a small goblet supporting the aphid
upright on its stalk, still held by the tip of the abdomen
and the hind feet (I). To liberate itself entirely the
insect must make a few more exertions (J), when, finally,
it pulls its legs and body from the grip of the drying
skin, and is at last a free young aphid (L).

The emergence from the egg and from the hatching
membrane is a critical period in the life of an aphid. The

PLATE 3

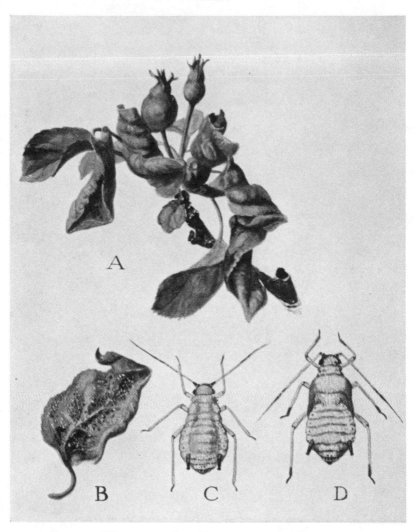

The rosy apple aphis (*Anuraphis roseus*)
A, apple leaves and young fruit distorted by the aphids; B, under surface of an infested leaf; C, immature wingless aphid (greatly enlarged); D, immature winged aphis

process may be completed in a few minutes, or it may take as long as half an hour, but if the feeble creature should be unable to free itself at last from the drying and contracting tissue, it remains a captive struggling in the grip of its embryonic vestment until it expires. The young aphid successfully delivered takes a few uncertain, staggering steps on its weak and colorless legs, and then complacently rests awhile; but after about twenty min-

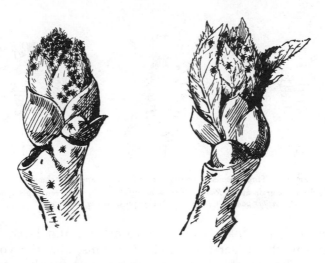

FIG. 94.   Young aphids on apple buds in spring

utes or half an hour it is able to walk in proper insect fashion, and it proceeds upward on its twig, a course sure eventually to lead it to a bud.

While the aphid eggs are hatching, or shortly thereafter, the apple buds are opening and unfolding their delicate, pale-green leaves, and from everywhere now the young aphids come swarming upon them, till the tips are often blackened by their numbers (Fig. 94). The hungry horde plunges into the hearts of the buds, and soon the new leaves are punctured with tiny beaks that rob them of their food; and the young foliage, upon which the tree depends for a proper start of its spring growth, is stunted and yellowed. Now is the time for the orchardist to spray if he has not already done so.

The entomologist, however, takes note that all the young aphids on the apple trees are not alike; perhaps there are three kinds of them in the orchard (Fig. 95), differing slightly, but enough to show that each belongs to a separate species. When the first buds infested are

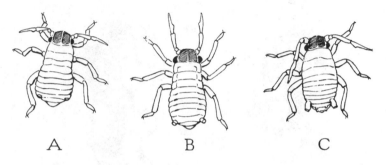

FIG. 95. Three species of young aphids found on apples in the spring
A, the apple-grain aphis, *Rhopalosiphum prunifoliae*. B, the green apple aphis, *Aphis pomi*. C, the rosy apple aphis, *Anuraphis roseus*

exhausted, the insects migrate to others, and later they spread to the larger leaves, the blossoms, and the young fruit. The aphids all grow rapidly, and in the course of two or three weeks they reach maturity.

The full-grown insects of this first generation, those produced from the winter eggs, are entirely wingless, and they are all females. But this state of affairs in no wise hinders the multiplication of the species, for these remarkable females are able of themselves to produce offspring (a faculty known as *parthenogenesis*), and furthermore, they do not lay eggs, but give birth to active young. Since they are destined to give rise to a long line of summer generations, they are known as the *stem mothers*.

One of the three aphid species of the apple buds is known as the *green apple aphis* (Fig. 95 B). During the early part of the season the individuals of this species are found particularly on the under surfaces of the apple leaves. They cause the infested leaves to curl and to become distorted in a characteristic manner (Fig. 96). The stem mothers (Fig. 97 A, B) begin giving birth to young (C) about twenty-four hours after reaching maturity, and any one of the mothers, during the course of her life of from ten to thirty days, may produce an average family of fifty or more daughters, for all her offspring are females, too. When these daughters grow up, however, none of them is exactly like their mother. They all have one more segment in each antenna; most of them are wingless (D), but many of them have wings—some, mere padlike stumps, but others well developed organs capable of flight (Fig. 97 E).

Both the wingless and the winged individuals of this second generation are also parthenogenetic, and they give birth to a third generation like themselves, including wingless, half-winged, and fully-winged forms, but with a greater proportion of the last. From now on there follows a

FIG. 96. Leaves of apple infested and distorted by the green apple aphis on under surfaces

large number of such generations continuing through the season. The winged forms fly from one tree to another, or to a distant orchard, and found new colonies. In summer, the green apple aphis is found principally on young shoots of the apple twigs, and on water sprouts growing in the orchard.

During the early part of the summer, the rate of production rapidly increases in the aphid colonies, and individuals of the summer generations sometimes give birth to young a week after they themselves were born. In the fall, however, the period of growth again is lengthened, and the families drop off in size; until the last females of the season produce each a scant half dozen young, though they may live to a much greater age than do the summer individuals.

The young summer aphids born as active insects are inclosed at birth in a tight-fitting, seamless, sleeveless, and legless tunic, as are those hatched from the winter eggs. Thus swathed, each emerges, rear end first, from the body of the mother, but is finally held fast by the face when it is nearly free. In this position, the embryonic bag splits over the head and contracts over the body of the young aphid to the tip of the abdomen, where it remains as a cap of shriveled membrane until it finally drops off or is pushed away by the feet. The infant, now vigorously kicking, is still held in the maternal grasp, and eventually liberates itself only after some rather violent struggling; but soon after it is free it walks away to find a feeding place among its companions on the leaf. The mother is but little concerned with the birth of her child, and she usually continues to feed during its delivery, though she may be somewhat annoyed by its kicking. The average summer female gives birth to two or three young aphids every day.

The succession of forms in the families is one of the most interesting phases of aphid life. Investigations have shown that the winged individuals are produced principally by wingless forms, and experiments have demonstrated that the occurrence of the winged forms is correlated with changes in the temperature, the food supply, and the duration of light. At a temperature around sixty-five degrees few winged individuals ever appear, but they are produced at temperatures either below or above this point. Likewise it has been found that when the food supply gives out through the drying

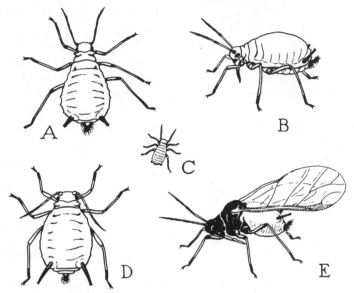

FIG. 97. The green apple aphis, *Aphis pomi*. A, B, adult stem mothers. C, a newly-born young of the summer forms. D, a wingless summer form. E, a winged summer form

of the leaves or by the crowding of the aphids on them, winged forms appear, thus making possible a migration to fresh feeding grounds. Then, too, certain chemical substances, particularly salts of magnesium, added to the water or wet sand in which are growing cuttings of plants infested with aphids, will cause an increase of winged forms in the insects subsequently born. This does not happen if the plants are rooted, but it shows that a change in the food *can* have an effect on wing production.

[165]

Finally, it has recently been shown experimentally by Dr. A. Franklin Shull that winged and wingless conditions in the potato aphis may be produced artificially by a variation in the relative amount of alternating light and darkness the aphids receive during each twenty-four hours. Shortening the illumination period to twelve hours or less results in a marked increase in the number of winged forms born of wingless parents. Continuous darkness, however, produces few winged offspring. Maximum results perhaps are obtained with eight hours of light. The effect of decreased light appears from Doctor Shull's experiments to be directly operative on the young from thirty-four to sixteen hours before birth, and it is not to be attributed to any physiological effect on the plant on which the insects are feeding.

It is evident, therefore, that various unfavorable local conditions may give rise to winged individuals in a colony of wingless aphids, thus enabling representatives of the colony to migrate in the chance of finding a more suitable place for the continuance of their line. The regular production of spring and fall migrants is brought about possibly by the shorter periods of daylight in the earlier and later parts of the season.

The final chapter of the aphid story opens in the fall and, like all last chapters done according to the rules, it contains the sequel to the plot and brings everything out right in the end.

All through the spring and summer the aphid colonies have consisted exclusively of virgin females, winged and wingless, that give birth to virgin females in ever-increasing numbers. A prosperous, self-supporting feminist dominion appears to be established. When summer's warmth, however, gives way to the chills of autumn, when the food supply begins to fail, the birth rate slackens and falls off steadily, until extermination seems to threaten. By the end of September conditions have reached a desperate state. October arrives, and the

surviving virgins give birth in forlorn hope to a brood
that must be destined for the end. But now, it appears,
another of those miraculous events that occur so frequently in the lives of insects has happened here, for the
members of this new brood are seen at once to be quite
different creatures from their parents. When they grow
up, it develops that they constitute a sexual generation,
composed of females and *males!* (Plate 2 A, B.)

Feminism is dethroned. The race is saved. The marriage instinct now is dominant, and if marital relations
in this new generation are pretty loose, the time is October, and there is much to be accomplished before winter
comes.

The sexual females differ from their virgin mothers
and grandmothers in being of darker green color and in
having a broadly pear-shaped body, widest near the end
(Plate 2 A). The males (B) are much smaller than the
females, their color is yellowish brown or brownish green,
and they have long spiderlike legs on which they actively
run about. Neither the males nor the females of the green
apple aphis have wings. Soon the females begin to produce, not active young, but eggs (D). The eggs are deposited most anywhere along the apple twigs, in crevices
where the bark is rough, and about the bases of the buds.
The newly-laid eggs are yellowish or greenish (D), but
they soon turn to green, then to dark green, and finally
become deep black (E). There are not many of them,
for each female lays only from one to a dozen; but it
is these eggs that are to remain on the trees through the
winter to produce the stem mothers of next spring, who
will start another cycle of aphid life repeating the history of that just closed.

The production of sexual forms in the fall in temperate
climates seems to have some immediate connection with
the lowered temperature, for in the tropics, it is said,
the aphid succession continues indefinitely through parthenogenetic females, and in most tropical species sexual

males and females are unknown. In the warmer regions of the West Coast of the United States, species that regularly produce males and females every fall in the East continue without a reversion to the sexual forms.

Of the other two species of apple aphids that infest the buds in the spring, one is known as the *rosy apple aphis* (Fig. 95 C). The name comes from the fact that the early summer individuals of this species have a waxy pink tint more or less spread over the ground color of green (Plate 3), though many of the adult stem mothers (Fig. 98 B) are of a deep purplish color. The early generations of the rosy aphis infest the leaves (Fig. 98 A, Plate 3 A) and the young fruit (Fig. 98 C, Plate 3 A), causing the former to curl up in tightly rolled spirals, and the latter to become dwarfed and distorted in form.

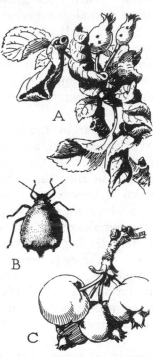

Fig. 98. The rosy apple aphis, *Anuraphis roseus*, on apple

A, a cluster of infested and distorted leaves. B, an adult stem mother. C, young apples dwarfed and distorted by the feeding of the aphids

The stem mothers of the rosy aphis give birth parthenogenetically to a second generation of females which are mostly wingless like their mothers; but in the next generation many individuals have wings. Several more generations now rapidly follow, all females; in fact, as with the green aphis, no males are produced till late in the season. The winged forms, however, appear in increasing numbers, and by the first of July almost all the individuals born have wings.

Heretofore, the species has remained on the apple trees, but now the winged ones are possessed with a desire for a change, a complete change both of scenery and of diet. They leave the apples, and when next discovered they are found to have established themselves in summer colonies on those common weeds known as plantains, and mostly on the narrow-leaved variety, the rib-grass, or English plantain (Fig. 99).

As soon as the migrants land upon the plantains they give birth to offspring quite unlike themselves or any of the preceding generations. These individuals are of a yellowish-green color and nearly all of them are wingless (Fig. 99). So well do they disguise their species that entomologists were a long time in discovering their identity. Generations of wingless yellow females now follow upon the plantain. But a weed is no fit place for the storage of winter eggs, so, with the advent of fall, winged forms again appear in abundance, and these migrate back to the apples. The fall migrants, however, are of two varieties: one is simply a winged female like the earlier migrants that came to the plantain from the apple, but the other is a winged male (Fig. 100 A). Both forms go back to the apple trees, and

FIG. 99. The rosy apple aphis on narrow-leaved plantain in summer; above, a wingless summer form (enlarged)

[169]

there the females give birth to a generation of wingless sexual females (B), which, when mature, mate with the males and produce the winter eggs.

The third of the aphid species that infest the spring buds of the apple is known as the *apple-grain aphis*, so called because, being a migratory species like the rosy

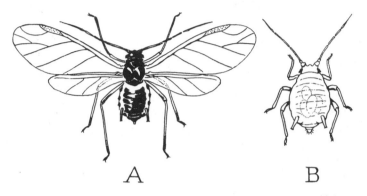

FIG. 100. The winged male (A) and the wingless sexual female (B) of the rosy apple aphis

aphis, it spends the summer upon the leaves of grains and grasses. The eggs of the apple-grain aphis are usually the first to hatch in the spring, and the young aphids of this species (Fig. 95 A) are distinguished by their very dark green color, which gives them a blackish appearance when massed upon the buds. Later they spread to the older leaves and to the petals of the apple blossoms, but on the whole their damage to the apple trees is less than that of either of the other two species. The summer history of the apple-grain aphids is similar to that of the rosy aphis, excepting that they make their summer home on grains and grasses instead of on plantains. In the fall, the winged female migrants (Plate 4) come back to the apple and there give birth to wingless sexual females, which are later sought out by the winged males.

It would be impossible here even to enumerate the

PLATE 4

The apple-grain aphis (*Rhopalosiphum prunifoliae*)

The winged form produced in the fall that migrates from the grain to the apple trees. (Enlarged 20 times)

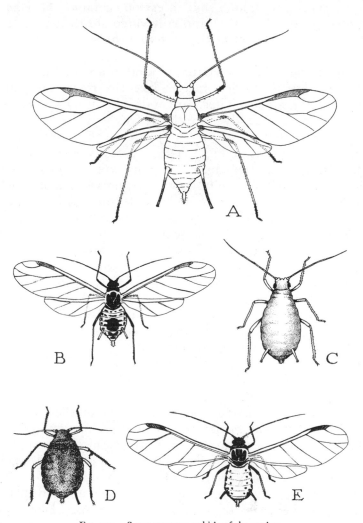

FIG. 101. Some common aphids of the garden

A, winged form of the potato aphis, *Illinoia solanifolii*, one of the largest of the garden aphids. B, winged form of the peach aphis, *Myzus persicae*, which infests peach trees and various garden plants. C, wingless form of the peach aphis. D, wingless form of the melon aphis, *Aphis gossypii*. E, winged form of the melon aphis

many species of aphids that infest our common field and garden plants (Fig. 101) and cultivated shrubs and trees, to say nothing of those that inhabit the weeds, the wild shrubbery, and the forest trees. Almost every natural group of plants has its particular kind of aphid, and many of them are migratory species like the rosy and grain aphis of the apple. There are root-inhabiting species as well as those that live on the leaves and stems. The *Phylloxera*, that pest of vineyards in California and France, is a root aphid. Those cottony masses that often appear on the apple twigs in late summer mark the presence of the woolly aphis, the individuals of which exude a fleecy covering of white waxy threads from their backs. The woolly aphis is more common on the roots of apple trees, being especially a pest of nursery stock, but it migrates to both the twigs and the roots of the apple from the elm, which is the home of its winter eggs.

An underground aphid of particular interest is one that lives on the roots of corn. We have seen that all aphids are much sought after by ants because of the honey dew they excrete, a substance greatly relished and prized by the ants. It is said that some ants protect groups of aphids on twigs by building earthen sheds over them; but the corn-root aphis owes its very existence to the ants. A species of ant that makes its nests in cornfields runs tunnels from the underground chambers of the nests to the bases of nearby cornstalks. In the fall the ants gather the winter eggs of the aphids from the corn roots and take them into their nests where they are protected from freezing during the winter. Then in the spring the ants bring the eggs up from the storage cellars and place them on the roots of various early weeds. Here the stem mothers hatch and give rise to several spring generations; but, as the new corn begins to sprout, the ants transfer many of the aphids to the corn roots, where they live and multiply during the summer and, in the fall, give birth to the sexual males and females, which produce the winter eggs. The eggs are again collected by the ants and carried to safety for the winter into the depths of their underground abodes. All this the ants do for the aphids in exchange for the honey dew they receive from them. The ants have so domesticated these corn-root aphids that the aphids would perish without their care. The farmer, therefore, who would rid his cornfield of the aphid pest, proceeds with extermination measures against the ants.

The crowded aphid colonies exposed on stems and leaves naturally form the happy hunting grounds for a

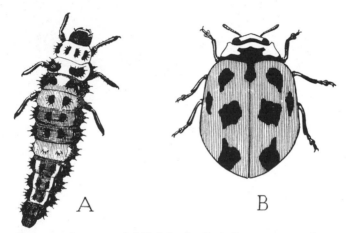

FIG. 102. A common ladybird beetle, *Coccinella novemnotata*, that feeds on aphids. (Enlarged 5 times)
A, the larva. B, the adult beetle

host of predacious insects. Here are thousands of soft-bodied creatures, all herded together, and each tethered to one spot by the bristles of its beak thrust deep into the tissues of the plant—a pot-hunter's paradise, truly. Consequently, the placid lives of the aphids have many interruptions, and vast numbers of the succulent creatures serve only as half-way stages in the food cycle of some other insect. The aphids have small powers of active defense. A pair of slender tubes, the *cornicles*, projecting from the rear end of the body, eject a sticky liquid which the aphids are said to smear on the faces of attacking insects; but the ruse at best probably does not give much protection. Parthenogenesis and large families are the principal policies by which the aphids insure their race against extinction.

The presence of "evil" in the world has always been a thorn for those who would preserve their faith in the idea of beneficence in nature. The irritation, however, is not

FIG. 103. The aphis-lion, feeding on an aphis held in its jaws

in the flesh but in a distorted growth of the mind, and consequently may be alleviated by a change of mental attitude. The thorn itself, however, is real and can not be explained away. Beneficence is not a part of the scheme by which plants and animals have attained through evolution their present conditions and relations. On the other hand, there are not good species and bad species; for every creature, including ourselves, is a thorn to some other, since each attacks a weaker that may contribute to its existence. There are many insects that destroy the aphids, but these are "enemies" of the aphids only in the sense that we are enemies of chickens and of cabbages, or of any other thing we kill for food or other purposes.

Recognizing, then, that evil, like everything else, is a matter of relativity and depends upon whose standpoint it is from which we take our view, it becomes only a pardonable bias in a writer if he views the subject from the standpoint of the heroes of his story. With this understanding we may note a few of the "enemies" of the aphids.

Everybody knows the "ladybirds," those little oval, hard-shelled beetles, usually of a dark red color with black spots on their rounded backs (Fig. 102 B). The female ladybirds, or better, lady-beetles, lay their orange-colored eggs in small groups stuck usually to the under surfaces of leaves (Fig. 132 B) and in the neighborhood of aphids. When the eggs hatch, they give forth, not ornate insects resembling lady-beetles, but blackish little beasts with thick bodies and six short legs. The young creatures at once seek out the aphids, for aphids are their natural food, and begin ruthlessly feeding upon them. As the young lady-beetles mature, they grow even uglier in form, some of them becoming conspicuously spiny, but their bodies are variegated with areas of brilliant color—red, blue, and yellow—the pattern differing according to the species. A common one is shown at A of Figure 102. When one of these miniature monsters becomes full-grown, it ceases its depredations on the aphid flocks, enters a period of quietude, and fixes the rear end of its body to a leaf by exuding a glue from the extremity of its abdomen. Then it sheds its skin, which shrinks down over the body and forms a spiny mat adhering to the leaf and supporting the former occupant by only the tip of the body (Fig. 132 E). With the shedding of the skin, the insect has changed from a *larva* to a *pupa*, and after a short time it will transform into a perfect lady-beetle like its father or mother.

Another little villian, a remarkably good imitation of a small dragon (Fig. 103), with long, curved, sicklelike jaws extending forward from the head, and a vicious tem-

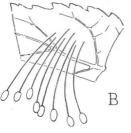

Fig. 104. The golden-eye, *Chrysopa*, the parent of the aphis-lion, and its eggs

A, the adult insect. B, a group of eggs supported on long thread-like stalks on the under surface of a leaf

perament to match, is also a common frequenter of the aphid colonies and levies a toll on the lives of the meek and helpless insects. This marauder is well named the *aphis-lion*. He is the larva of a gentle, harmless creature with large pale-green lacy wings and brilliant golden eyes (Fig. 104 A). The parent females show a remarkable prescience of the nature of their offspring, for they support their eggs on the tips of long threadlike stalks, usually attached to the under surfaces of leaves (B). The device seems to be a scheme for preventing the first of the greedy brood that will hatch from devouring its own brothers and sisters still in their eggs.

Wherever the aphids are crowded there is almost sure to be seen crawling among them soft grayish or green wormlike creatures, mostly less than a quarter of an inch in length. The body is legless and tapers to the forward end,

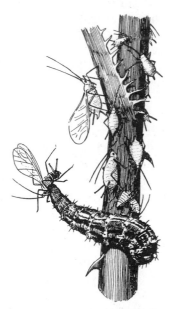

Fig. 105. A larva of a syrphus fly feeding on aphids

[176]

which has no distinct head but from which is protruded and retracted a pair of strong, dark hooks. Watch one of these things as it creeps upon an unsuspecting aphid; with a quick movement of the outstretched forward end of the body it makes a swing at the fated insect, grabs it with the extended hooks, swings it aloft kicking and struggling, and relentlessly sucks the juices from its body (Fig. 105). Then with a toss it flings the shrunken skin aside, and repeats the attack on another aphid. This heartless blood-sucker is a maggot, the larva of a fly (Fig. 106) belonging to a family called the Syrphidae. The adult flies of this family are entirely harmless, though

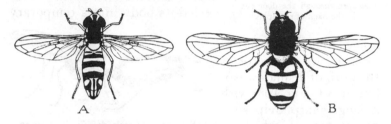

Fig. 106. Two common species of syrphus flies whose larvae feed on aphids.
(Enlarged about 3¼ times)
A, *Allograpta obliqua*.   B, *Syrphus americana*

some of them look like bees, but the females of those species whose maggots feed on aphids know the habits of their offspring and place their eggs on the leaves where aphids are feeding. One of them may be seen hovering near a well-infested leaf. Suddenly she darts toward the leaf and then as quickly is off again; but in the moment of passing, an egg has been stuck to the surface right in the midst of the feeding insects. Here it hatches where the young maggot will find its prey close at hand.

In addition to these predacious creatures that openly and honestly attack their victims and eat them alive, the aphids have other enemies with more insidious methods of procedure. If you look over the aphid-infested leaves

FIG. 107. A dead potato aphis that has contained a parasite, which when adult escaped through the door cut in the back of the aphis

on almost any plant, you will most likely note here and there a much swollen aphid of a brownish color. Closer examination reveals that such individuals are dead, and many of them have a large round hole in the back, perhaps with a lid standing up from one edge like a trap door (Fig. 107). These aphids have not died natural deaths; each has been made the involuntary host of another insect that converted its body into a temporary

home. The guest that so ravishes its protector is the grub of a small wasplike insect (Fig. 108) with a long, sharp ovipositor by means of which it thrusts an egg into the body of a living aphid

FIG. 109. A female *Aphidius* inserting an egg into the body of a living aphid, where the egg hatches; the larva grows to maturity by feeding in the tissues of the aphis. (From Webster)

(Fig. 109). Here the egg hatches and the young grub feeds on the juices of the aphid until it is itself full-grown, by which time the aphid is exhausted and dead. Then the grub slits open the lower wall of the hollow corpse and spins a web between the lips of the opening and against the surface of the leaf below, which attaches the aphid shell to the support. Thus secured, the grub proceeds to give

FIG. 108. *Aphidius*, a common small wasplike parasite of aphids

[178]

its gruesome chamber a lining of silk web; which done, it lies down to rest and soon changes to a pupa. After a short time it again transforms, this time into the adult of its species, and the latter cuts with its jaws the hole in the back of the aphid and emerges.

In other cases, the dead aphid does not rest flat on the leaf but is elevated on a small mound (Fig. 110 A). Such victims have been inhabited by the grub of a related species, which, when full-grown, spins a flat cocoon beneath the dead body of its host, and in this inclosure undergoes its transformation. The adult insect then cuts a door in the side of the cocoon (B), through which it makes its exit.

Insects that usurp the bodies of other insects for their own purposes are called *parasites*. Parasites are the

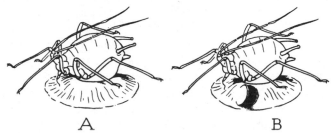

A   B

FIG. 110.   Aphids parasitized by a parasite that makes a cocoon beneath the body of the aphis, where it changes to a pupa and, when adult, emerges through a door cut in the side of the cocoon

worst enemies that insects have to contend against; but really they do not contend against them in most cases, except in the way characteristic of insects, which is to insure themselves against extermination by the number of their offspring. The aphid colonies are often, however, greatly depleted during a season favorable to the predacious and parasitic insects that attack them; but no species is ever annihilated by its enemies, for this would mean starvation to the next year's brood of the latter. The laws of compensation usually maintain a balance in nature between the procreative and the destructive forces.

The insect parasites and predators of other insects in general comprise a class of insects that are most beneficial to us by reason of their large-scale destruction of species injurious to our crops. But, unfortunately, parasites as a class do not respect our classification of other creatures into harmful and useful species. Even as some predator is stalking its prey, another insect is likely to be shadowing it, awaiting the chance to inject into its body the egg which will mean finally death to the destroyer. Immature insects are often found in a sluggish or half-alive condition, and an internal examination of their bodies usually reveals that they are occupied by one or more parasitic larvae. A larva of any of the lady-beetles, for example, is frequently seen attached to a leaf for pupation (Fig. 111), which, instead of transforming to a pupa, remains inert and soon becomes a lifeless form, though still adhering to the leaf and bent in the attitude that the pupa would assume. In a short time there issues through the dried skin a parasite, giving evidence of the fate that has befallen the unfortunate larva; even if the usurper is not seen, the exit hole in the larval skin bears witness to his former occupancy and escape.

And the parasites themselves, do they lead unmolested lives? Are they the final arbiters of life and death in the insect world? If you are fortunate sometime while studying aphids out-of-doors, you may see a tiny black mite, no bigger than the smallest gnat, hovering about an infested plant or darting uncertainly from one leaf to another, with the air of searching for something but not knowing just where to look. You would probably suspect the intruder of being a parasite seeking a chance to place an egg in the body of an aphid; but here she hovers over a group of fat lice without selecting a victim, then perhaps alights and runs about on the leaf nervously and intensely eager, still finding nothing to her choice. Her senses must be dull, indeed, if it is aphids that she wants. Do not lose sight of her, however, for her attitude has changed; now she certainly has her eye upon something that holds her attention, but the object is nothing other than one of those swollen parasitized aphids. Yet she excitedly runs up to it, feels it, grasps it, mounts upon it, examines it all over. Evidently she is satisfied. She dismounts, turns about, backs her abdomen against the inflated mummy; now out comes the swordlike ovipositor, and with a thrust it is sunken into the already parasitized aphid. Two minutes later her business is ended, the ovipositor is withdrawn, once more sheathed, and the insect is off and away.

This tiny creature is a *hyperparasite*, which is to say, a parasite of a parasite. In the act just witnessed she, too, has thrust an egg into the aphid, but the grub that will hatch from it will devour the parasitic occupant that is already in possession of the aphid's skin.

FIG. 111. A parasitized larva of a lady-bird beetle, and one of the parasites

The larva of the beetle has attached itself to a leaf preparatory to pupation, but has not changed to a pupa because of the parasites within it. Above, one of the parasites, which escaped from the beetle larva through a hole it cut in the skin of the latter

There are also parasites of hyperparasites, but the series does not go on *"ad infinitum"* as in the old rhyme, for the limitation of size must intervene.

CHAPTER VII

THE PERIODICAL CICADA

It is to be observed, in most of our human affairs, that we give greatest acclaim to the spectacular, and, furthermore, that when once a hero has delivered the great thrill, all his acts of everyday life acquire headline values. Thus a biographer may run on at great length about the petty details in the life of some great person, knowing well that the public, under the spell of hero worship, will read with avidity of things that would make but the dullest platitudes if told of any undistinguished mortal. Therefore, in the following history of our famous insect, universally known as the "seventeen-year locust," the writer does not hesitate to insert matter that would be dry and tedious if given in connection with a commonplace creature.

Most unfortunate it is, now, that we are compelled to divest our hero of his long-worn epithet of "seventeen-year locust," and to present him in the disguise of his true patronymic, which is *cicada* (pronounced *sĭ-ka'-da*). In a scientific book, however, we must have full respect for the proprieties of nomenclature; and since, as already explained in Chapter I, the name "locust" belongs to the grasshopper, we can not continue to designate a cicada by this term, for so doing would but propagate confusion. Moreover, even the praenomen of "seventeen-year" is misleading, for some of the members of the species have thirteen-year lives. Entomologists, therefore, have rechristened the "seventeen-year locust" the *periodical cicada*.

The cicada family, the Cicadidae, includes many species

of cicadas in both the New World and the Old, and some of them are more familiar, at least by sound, than our periodical cicada, because not only are the males notoriously musical, but they are to be heard every year (Fig. 112). The cicadas of southern Europe were highly esteemed by the ancient Greeks and Romans for their song, and they were often kept in cages to furnish entertainment

FIG. 112. One of the common annual cicadas whose loud song is heard every year through the later part of the summer

with their music. The Greeks called the cicada *tettix*, and Aesop, who always found the weak spot in everybody's character, wrote a fable about the tettix and the ant, in which the tettix, or cicada, after having sung all summer, asked a bite of food from the ant when the chill winds of coming winter found him unprovisioned. But the practical ant heartlessly replied, "Well, now you can dance." This is an unjust piece of satire because the moral is drawn to the disparagement of the cicada. Human musicians have learned their lesson, however, and sign their contracts with the box-office management in advance.

INSECTS

In the United States there are numerous species of "annual" cicadas, so called because they appear every year, but their life histories are not actually known in most cases. These species are called "locusts," "harvest flies," and "dog-day cicadas" (Fig. 112). They are the insects that sit in the trees during the latter half of summer and make those long shrill sounds that seem to be the natural accompaniment of hot weather. Some give a rising and falling inflection to their song, which resembles *zwing, zwing, zwing, zwing,* (repeated in a long series); others make a vibratory rattling sound; and still others utter just a continuous whistling buzz.

During the interval between the times of the appearance of the adult cicadas, the insects live underground. The periodical cicada comprises two races, one of which lives in its subterranean abodes for most of seventeen years, the other for most of thirteen years. Both races inhabit the eastern part of the United States, but the longer-lived race is northern, and the other southern, though their territories overlap. Most of our familiar insects complete their life cycle in a single year, and many of them produce two or more generations every season. For this reason we marvel at the long life of the periodical cicada. Yet there are other common insects that normally require two or three years to reach maturity, and certain beetles have been known to live for twenty years or more in an immature stage, though under conditions adverse for transforming to the adult.

Throughout the period of their underground life the cicadas have a form quite different from that which they take on when they leave the earth to spend a brief period in the trees. The form of the young periodical cicada at the time it is ready to emerge from the ground is shown in Plate 5. It will be seen that it suggests one of those familiar shells so often found clinging to the trunk of a tree or the side of a post. These shells, in fact, are the empty skins of young cicadas that have discarded their

PLATE 5

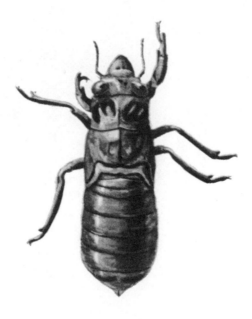

The mature nymph of the periodical
cicada in the form in which it leaves the
ground to transform to the winged adult
after a subterranean life of nearly
seventeen years

earthly form for that of a winged insect of the upper world and sunshine, though the skins ordinarily seen are those of the annual species.

The cicada undergoes a striking transformation from the young to the adult, but it does so directly and not by means of an intervening stage, or pupa. The young of an insect that transforms directly is termed a *nymph* by most American entomologists. The last nymphal stage is sometimes called a "pupa," but it is not properly so designated.

The life of the periodical cicada stirs our imagination as that of no other insect does. For years we do not see the creatures, and then a springtime comes when countless thousands of them issue from the earth, undergo their transformation, and swarm into the trees. Now, for several weeks, the very air seems swayed with the monotonous rhythm of their song, while the business of mating and egg-laying goes rapidly on; and soon the twigs of trees and shrubs are everywhere scarred with slits and punctures where the eggs have been inserted. In a few weeks the noisy multitude is gone, but for the rest of the season the trees bear witness to the busy throng that so briefly inhabited them by a spotting of their foliage with masses of brown and dying leaves where the punctured stems have broken in the wind. The young cicadas that hatch from the eggs later in the summer silently drop to the earth and hastily bury themselves beneath the surface. Here they live in solitude, seldom observed by creatures of the upper world, through the long period of their adolescent years, only to enjoy at the end a few brief weeks of life in the open air in the fellowship of their kind.

THE NYMPHS

Of the underground life of the periodical cicada we still know very little. The fullest account of the history of this species is that given by Dr. C. L. Marlatt in his

Bulletin, *The Periodical Cicada*, published by the United States Bureau of Entomology in 1907. Doctor Marlatt describes six immature stages of the periodical cicada between the egg and the adult.

The young cicada that first enters the ground is a minute, soft-bodied, pale-skinned creature about a twelfth of an inch in length (Fig. 126). The body is cylindrical and is supported on two pairs of legs, the front legs being the digging organs; the somewhat elongate head bears a pair of small dark eyes and two slender, jointed antennae. At no stage has the cicada jaws like those of the grasshopper; it is a sucking insect, related to the aphids, and is provided with a beak arising from the under surface of the head, but when not in use the beak is turned backward between the bases of the front legs. Throughout the period of its underground life, the cicada subsists on the sap of roots.

During more than a year the young cicada retains approximately the form it has at hatching, though the body changes somewhat in shape, principally by an increase in the size of the abdomen (Fig. 113). According to Doctor Marlatt, a nymph of the seventeen-year race first

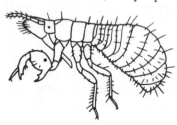

sheds its skin, or molts, sometime during the first two or three months of the second year of its life.

In its second stage it becomes a little larger and is marked by a change in the structure of the front legs, the terminal foot part of each being reduced to a mere spur and the fourth section being developed into

FIG. 113. Nymph of the periodical cicada in the first stage, about 18 months old, enlarged 15 times. (From Marlatt)

a strong, sharp-pointed pick which forms a more efficient organ for digging. The second stage lasts nearly two years; then the creature molts again and enters its third stage, which is about a year in length. In the fourth stage, which lasts perhaps three or four years, the nymph (Fig. 114) shows distinct wing pads on the two wing-bearing segments of the thorax. In the fifth stage the insect, sometimes now called a "pupa," takes on the form it has when it finally emerges from the earth; its front feet are restored and its wing pads are well developed, but it has entirely lost its small nymphal eyes. Once more, before its long underground sen-

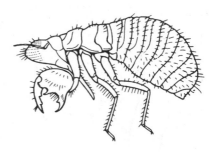

Fig. 114. Nymph of the periodical cicada in the fourth stage, about 12 years old, enlarged 2¾ times. (From Marlatt)

tence is up, the nymph molts, and enters the sixth and last stage of its subterranean life. When mature (Plate 5) it is about an inch and a quarter in length, thick-bodied, and brown in color; it appears to have a pair of bright-red eyes on the head, but these are the eyes of the adult inside showing through the nymphal skin.

According to the investigations of Doctor Marlatt, the nymphs of the periodical cicada do not ordinarily burrow into the earth below two feet, and most of them are to be found at depths varying from eight to eighteen inches. However, there are reports of their having been discovered ten feet beneath the surface, and they have been known to emerge from the floors of cellars at the time of transformation to the adult stage. There is no evidence that the insects, even when present in great numbers in the earth, do any appreciable damage to the vegetation on the roots of which they feed.

Some time before the mature nymphs emerge from the ground, probably in April of the last year of their lives, the insects come up from their subterranean burrows and construct a chamber of varying depth just below the

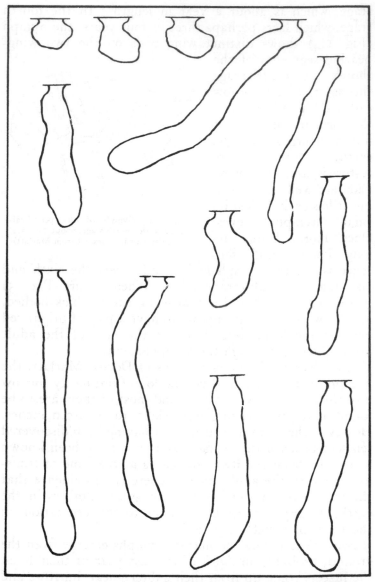

FIG. 115. Outlines of plaster casts of underground resting chambers of the mature nymph of the periodical cicada (about one-half natural size)

surface. A good idea of the size and shape of these chambers may be obtained by filling the opened holes with a mixture of plaster of Paris in water, letting the plaster harden, and then digging up the casts. Figure 115 shows casts of a number of chambers made in this way. Some, it is seen, are mere cups about an inch in depth, but most of them are long and narrow, descending several inches into the ground, the longest being six inches or more in depth. The width is usually about five-eighths of an inch. All the chambers have a distinct enlargement at the bottom, and most of them are slightly widened at the top. The upper wall of each is separated from the surface by a layer of undisturbed soil about half an inch in thickness, which is not broken until the insect is ready to emerge.

The shafts are seldom straight, their courses being more or less tortuous and inclined to the surface, as the miner had to avoid roots and stones obstructing the vertical path. The interior contains no débris of any kind, and the walls are smooth and compact. Below each chamber there is always evidence of a narrower burrow going irregularly downward into the earth, but this tunnel is filled to the chamber floor with black granular earth. The burrows examined by the writer near Washington in 1919 were dug through compact red clay, and the lower tunnels here made a distinct black path through the red of the surrounding clay, where some could be followed for a considerable distance. The black color of the earth filling the tunnels was possibly due to an admixture of fecal matter.

The chambers, as we have noted, are closed at the top until the cicada is ready to emerge. The largest chambers are many times the bulk of the nymph in volume, and it becomes, then, a question as to what the insect does with the material it removed in making a hole of such size. It seems improbable that it could have been carried down into the lower tunnel, for this would be filled with its own débris. The insects themselves will give an answer to the question if several of them are placed in glass tubes and covered with earth; but, to understand the cicada's technique, we must first study the mechanism of its digging tools, the front legs.

The front leg of a mature cicada nymph (Fig. 116 A) is composed of the same parts as any other of its legs. The third segment from the base, which is the *femur* (*F*), is large and swollen, and has a pair of strong spines and a comb of smaller ones projecting from its lower edge. The next segment is the *tibia* (*Tb*). It is curved and terminates in a strong recurved point (B). Finally, attached to the inner surface of the tibia, well up from its terminal point, is the slender *tarsus* (*Tar*). The tarsus can be extended beyond the

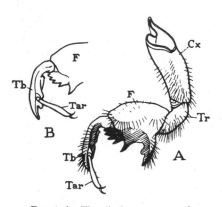

FIG. 116. The digging organ, or front leg, of the mature cicada nymph

A, right leg, inner surface (4 times natural size). B, the tarsus (*Tar*) bent inward at right angles to the tibia (*Tb*), the position in which it is used as a rake

Cx, basal joint or coxa; *Tr*, trochanter; *F*, femur; *Tb*, tibia; *Tar*, tarsus, with two terminal claws

tibial point when the insect is walking or climbing, but can also be turned inward at a right angle to the latter, as shown at B, or bent back against the inner surface of the tibia.

Let us now return to the insects in the earth-filled tubes, where they are industriously at work. It will be seen that they are using the curved, sharp-pointed tibiae as picks with which to loosen the earth, the tarsi being turned back and out of the way. The two legs, working alternately, soon accumulate a small mass of loosened material in front of the insect's body. Now there is a cessation of digging and the tarsi are turned forward at right angles to the tibiae to serve as rakes (Fig. 116 B). The mass of earth pellets is scraped in toward the body, and—here comes the important part, the cicada's special technique—the little pile of rakings is grasped by one front leg between the tibia and the femur (Fig. 116 A, *Tb* and *F*), the former closing up against the spiny margin of the latter, the leg strikes forcibly outward, and the mass of loosened earth is pushed back into the surrounding earth. The process is repeated, first with one leg, then with the other. The miner looks like a pugilist training on a punching bag. Now and then the worker stops and rubs his legs over the protruding front of the head to clean them on the rows of bristles which cover each side of the face. Then he proceeds again, clawing, raking, gathering up the loosened particles, thrusting them back into the wall of the growing chamber. His back is firmly pressed against the opposite side of the cavity, the middle legs are bent forward until their knees are almost against the bases of the front legs, their tibiae lying along the wing pads. The hind legs keep a normal position, though held close against the sides of the body.

From what we know of the cicada's spring habits underground, we can infer that the nymphs construct their chambers on their arrival near the surface during April, and that, when the chambers are completed, the insects wait within for the signal to emerge and transform into the adult. Then they break through the thin caps at the surface and come out. It would be difficult to explain how they know when they are so near the top of the ground, and why some construct ample chambers several inches deep while others make mere cells scarcely larger than their bodies. Do they burrow upward till the pressure tells them that the surface is only a quarter of an inch or so away, and then widen the débris-filled tunnel downward? Evidently not, because the chamber walls are made of clean, compacted clay in which there is no admixture of the blackened contents of the burrows. It is unlikely, too, that they base their judgments on a sense of temperature, because their acts are not regulated by the nature of the season, which, if early or late, would fool them in their calculations.

Early in the spring, before the proper emergence season, cicada nymphs are often found beneath logs and stones. This is to be expected, for, to the ascending insect, something impenetrable has blocked the way, and there is nothing to tell it that it has already reached the level of the surface.

A more curious thing, often observed in some localities, is that the insects sometimes continue their chambers up above the surface of the ground within closed turrets of mud from two to several inches in height (Fig. 117). At certain places these cicada "huts" have been reported as occurring in great numbers; and it has been supposed that they may be built wherever there is something about the nature of the soil that the insects do not like, the earth being perhaps too damp, for they are frequently found where the ground

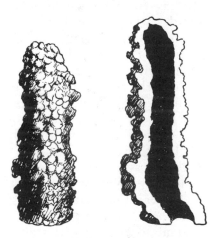

FIG. 117. Earthen turrets sometimes erected by the nymphs of the periodical cicada as continuations from their underground chambers. One cut open showing the tubular cavity within. (From photograph by Marlatt)

is unusually wet. On the other hand, the turrets have been observed in dry situations as well, and towers and holes flush with the surface frequently occur intermingled. The writer has had no opportunity of studying the cicada turrets, but a most interesting description of them is given

PLATE 6

The cicada just after emergence from the nymphal skin. (Enlarged two-thirds)

THE PERIODICAL CICADA

by Dr. J. A. Lintner in his *Twelfth Report on the Insects of New York*, published in 1897. Dr. Lintner says the turrets are constructed by the nymphs with soft pellets of clay or mud brought up from below and firmly pressed into place, and he records an observation on a nymph caught at work with a pellet of mud in its claws. We may infer, then, that the cicada's style of work as a mason is only a modification of its working methods as a miner, but it appears that no one has yet actually watched the construction of one of the turrets. At emergence time the towers are opened at the top and the insects come forth as they would from an ordinary chamber beneath the level of the ground.

THE TRANSFORMATION

The period of emergence for most of the cicadas of the northern, or seventeen-year, race is the latter part of May. The time of their appearance over large areas is much more nearly uniform than with most other insects, which show a wide variation according to temperature as determined by the season, the elevation, and the latitude. Nevertheless, observations in different localities show that the cicada, too, is influenced by these conditions. In the South, members of the thirteen-year race may emerge even a month earlier, the first individuals of the southernmost broods appearing in the latter part of April.

By some feeling of impending change the mature nymph, waiting in its chamber, knows when the time of transformation is at hand. Somehow nature regulates the event so that it will happen in the evening, but, once the hour has come, no time is to be lost. The nymph must break out of its cell, find a suitable molting site and one in accord with the traditions of its race, and there fix itself by a firm grip of the tarsal claws. At the beginning of the principal emergence period large numbers of the insects come out of their chambers as early as five o'clock in the afternoon; but after the rush of the first few days not many appear before dusk.

It is difficult to catch a nymph in the very act of making its exit from the ground, and apparently no observations have been recorded on the manner of its leaving. Do the insects leisurely open their doors some time in advance of their actual need and wait below till the proper hour, or do they break through the thin caps of earth and emerge at once? Digging up many open chambers revealed a living nymph in only one. Another issued from one of several dozen holes filled with liquid plaster for obtaining casts. Add to this the fact that great numbers of fresh holes are to be seen every morning during the emergence season, and the evidence would appear to indicate that the insects open their doors in the evening and come out at once. Only one chamber was found in the daytime partly opened.

If the insects are elusive and wary of being spied upon as they make their début into the upper world, a witness of their subsequent behavior does not embarrass them at all. However, events are imminent; there is no time to waste. The crawling insects head for any upright object within their range of vision—a tree is the ideal goal if it can be attained, and since the creatures were born in trees there is likely to be one near by. Yet it frequently happens that trees in which many were hatched have been since cut down, in which case the returning pilgrims must make a longer journey perhaps than anticipated. But the transformation can not be delayed; if a tree is not accessible, a bush or a weed, a post, a telegraph pole, or a blade of grass will do. On the trees some get only so far as the trunk, others attain the branches, but the mob gets out upon the leaves. Though thousands emerge almost simultaneously, they have not all been timed alike. Some have but a few minutes to spare, others can travel about for an hour or so before anything happens.

THE PERIODICAL CICADA

The external phase of transformation, more strictly the shedding of the last nymphal skin, has been many times observed. It is nothing more than what all insects do. But the cicada is notorious because it does the thing in such a spectacular way, almost courting publicity where most insects are shy and retiring. As a consequence the cicada is famous; the others are known only to prying entomologists.

Let us suppose now that our crawling nymph has reached a place that suits it, say on the trunk of a tree, or better still on a piece of branch provided for it and taken into a lighted room where its doings can be more clearly observed. Though the insects choose the evening for emergence, they are not bashful at all about changing their clothes in the glare of artificial light. The progress of this performance is illustrated by Figure 118. The first drawing shows the nymph still creeping upward; but in the next (2) it has come to rest and is cleaning its front feet and claws on the brushes of its face, just as did those confined to the glass tubes to give a demonstration of their digging methods. The front feet done, the hind ones are next attended to. First one and then the other is slowly flexed and then straightened backward (3) while the foot scrapes over the side of the abdomen. Several times these acts are repeated calmly and deliberately, for it is an important thing that the claws be well freed from any particles of dry earth that might impair their grip on the support. At last the toilet is completed, though the middle feet are always neglected, and the insect feels about on the twig, grasping now here, now there, till its claws take a firm hold on the bark. At the same time it sways the body gently from side to side as if trying to settle comfortably for the next act.

Thirty-five minutes may be consumed in the above preliminaries and there is next a ten-minute interval of quietude before the real show begins. Then suddenly

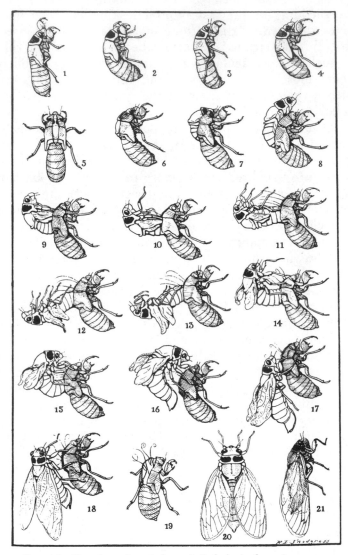

FIG. 118. Transformation of the periodical cicada from the mature
nymph to the adult

the insect humps its back (*4*), the skin splits along the midline of the thorax (*5*), the rupture extending forward over the top of the head and rearward into the first segment of the abdomen. A creamy white back, stamped with two large jet-black spots, now bulges out (*6, 7*); next comes a head with two brilliant red eyes (*8*); this is followed by the front part of a body (*9*) which bends backward and pulls out legs and bases of wings. Soon one leg is free (*10*), then four legs (*11*), while four long, glistening white threads pull out of the body of the issuing creature but remain attached to the empty shell. These are the linings of the thoracic air tubes being shed with the nymphal skin. Now the body hangs back down, when all the legs come free (*12*), and now it sags perilously (*13*) as the wings begin to expand and visibly lengthen.

Here another rest intervenes; perhaps twenty-five minutes may elapse, while the soft new creature, like an inverted gargoyle supported only by the rear end of its body, hangs motionless far out from the split in the back of the shell. Now we understand why the nymph took such pains to get a firm anchorage, for, should the dead claws give way at this critical stage, the resulting fall most probably would prove fatal.

The next act begins abruptly. The gargoyle moves again, bends its body upward (*14*), grasps the head and shoulders of the slough (*15*), and pulls the rear parts of its body free from the gaping skin (*16*). The body straightens and hangs downward (*17*). At last we behold the free imago, not yet mature but rapidly assuming the characters of an adult cicada. The new creature hangs for a while from the discarded shell-like skin, clinging by the front and middle legs, sometimes by the first alone; the hind ones spread out sideways or bend against the body, rarely grasping the skin. The wings continue to unfold and lengthen, finally hang flat, fully formed, but soft and white (*18*). Here the creature usually becomes restless, leaves the empty skin (*19*), and takes up a new position several inches away (*20*).

At this stage the cicada is strangely beautiful. Its creamy-yellow paleness, intensified by the great black patches just behind the head and relieved by the pearly flesh tint of the mesothoracic shield, its shining red eyes, and the milky, semitransparent wings with deep chrome on their bases make a unique impression on the mind. There is a look of unreality about the thing, which out of doors (Plate 6) becomes a ghostlike vision against the night. But, even as we watch, the color changes; the unearthly paleness is suffused with bluish gray, which deepens to blackish gray; the wings flutter, fold against the back, and the spell is broken—an insect sits in the place of the vanished specter.

The rest is commonplace. The colors deepen, the grays become blackish and then black, and after a few hours the creature has all the characters of a fully matured cicada. Early the next morning it is fluttering about, restless to be off with its mates to the woods.

The time consumed by the entire performance, from the splitting of the skin (Fig. 118, *5*) to the folding of the wings above the back (*21*), varies with different individuals, observed at the same time and under the same conditions, from forty-five minutes to one hour and twelve minutes. Most of the insects have issued from the nymphal skins before eleven o'clock at night, but occasionally a straggler may be seen in the last act as late as nine o'clock the following morning—probably a belated arrival who overslept the night before.

Thus, to the eye, the burrowing and crawling creature of the earth becomes transfigured to a creature of the air; yet the visible change is mostly but the final escape of the mature insect from the skin of its preceding stage. Aside from a few last adjustments and the expansion of the wings, the real change has been in progress within the nymphal skin perhaps for years. We do not truly witness

PLATE 7

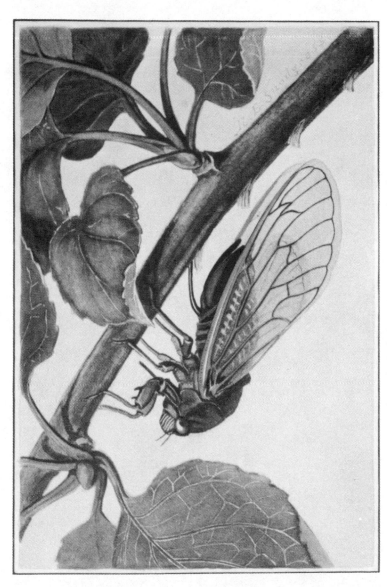

The periodical cicada (*Magicicada septendecim*)

A female inserting eggs with her ovipositor into the under surface of an apple twig. (Enlarged two times)

the transformation; we see only the throwing off of the shell that concealed it, as the circus performer strips off the costume of the clown and appears already dressed in that of the accomplished acrobat.

THE ADULTS

The adult cicada bears the stamp of individuality. In form he does not closely resemble any of our everyday insects, and he has a personality all his own; he impresses us as a "distinguished foreigner in our midst." The body of the periodical cicada is thick-set (Fig. 119), the face is bulging, the forehead is wide, with the eyes set out prominently on each side; from the under side of the head the short, strong beak projects downward and backward between the bases of the front legs. The colors are distinctive but not striking. The back is plain black (Plate 7); the eyes are bright red; the wings are shiny transparent amber with strongly marked orange-red veins; the legs and beak are reddish, and there are bands of the same color on the rings of the abdomen. Each front wing is branded near the tip with a conspicuous dark-brown W.

With both the seventeen-year race and the thirteen-year race of the periodical cicada there is associated a small cicada, which, however, differs so little except in size from the others (Fig. 119) that entomologists generally regard it as a mere variety of the larger form, the latter always including by far the greater number of individuals in any brood.

The male cicada has a pair of large drumheads beneath the bases of the wings on the front end of the abdomen (Fig. 120, *Tm*). These are the instruments by which he produces his music, and we will give them more attention presently. The female cicada has no drums nor other sound-making organs; she is voiceless, and must keep silence no matter how much her noisy mate may disturb her peace. The chief distinction of the female is her ovipositor, a long, swordlike instrument used for inserting the eggs into the twigs of trees and bushes. Ordinarily the ovipositor is kept in a sheath beneath the rear half of the abdomen, but when in use it can be turned downward and forward by a hinge at its base (Plate 7). The ovipositor consists of two lateral blades, and a guide-rail above. The blades excavate a cavity in the wood into which the eggs are passed through the space between the blades.

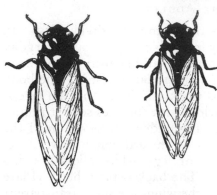

FIG. 119. Males of the large and small form of the periodical cicada (natural size)

It was formerly supposed that the periodical cicada takes no food during the brief time of its adult life, but we know from the observations of Mr. W. T. Davis, Dr. A. L. Quaintance, and others and from a study of the stomach contents made by the writer that the insects do feed abundantly by sucking the sap from the trees on which they live. The cicada, being a near relative of the aphids, has also, as we have already noted, a piercing and sucking beak by which it punctures the plant tissues and draws the sap up to its mouth. Unlike the other sucking insects that infest plants,

FIG. 120. A male of the periodical cicada with the wings spread, showing the ribbed sound-producing organs, or tympana (*Tm*), on the base of the abdomen

[200]

THE PERIODICAL CICADA

however, the cicadas cause no visible damage to the trees by their feeding. Perhaps this is because their attack lasts such a short time and comes at a season when the trees are at their fullest vigor.

The details of the head structure of the cicada and the exposed part of the beak are shown in Figure 121, which gives in side view the head of a fully matured adult, detached from the body by the torn neck membrane (*NMb*), with the beak (*Bk*) extending downward and backward below. The large eyes (*E*) project from the sides of the upper part of the head. The face is covered by a large protruding, striated plate (*Clp*). The cheek regions are formed by a long plate (*Ge*) on each side below the eyes; and between each cheek plate and the striated facial plate is partly concealed a narrower plate (*Md*). The cicada has no jaws. Its true mouth is shut in between the large flap (*AClp*), below the striated facial plate, and the base of the beak.

If the outer parts of the head about the mouth can be separated, there will be seen within them some other very important parts ordinarily hidden from view. In a specimen that has been killed in the act of emerging from the nymphal skin, when it is still soft, the outer parts are easily separated, exposing the structures shown at B of the same figure.

It is now to be seen (Fig. 121 B) that the beak consists of a long troughlike appendage (*Lb*) suspended from beneath the back part of the head, having a deep groove on its front surface in which are normally ensheathed two pairs of slender bristles (*MdB*, *MxB*), of which only the two of the left side are shown in the figure. In front of the bases of the bristles there is exposed a large tongue-like organ which is the hypopharynx (*Hphy*). Between this tongue and the flap hanging from the front of the face is the wide-open mouth (*Mth*), the roof of which (*e*) bulges downward and almost fills the mouth cavity. The way in which the cicada obtains its liquid food de-

pends upon the finer structure and the mechanism of
the parts before us.

Each one of the second pair of bristles has a furrow
along the entire length of its inner surface, and the two

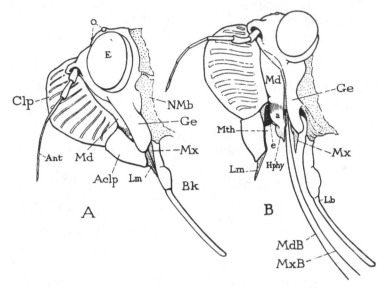

Fig. 121. The structure of the head and sucking beak of the adult cicada
A, the head in side view with the beak (*Bk*) in natural position

B, the head of an immature adult: the mouth (*Mth*) opened, exposing the roof
(*e*) of the sucking pump (see fig. 122), and the tonguelike hypopharynx (*Hphy*);
the parts of the beak separated, showing that it is composed of the labium
(*Lb*), inclosing normally two pairs of long slender bristles (*MdB*, *MxB*, only
one of each pair shown)

a, bridge between base of mandibular plate (*Md*) and hypopharynx (*Hphy*);
Aclp, anteclypeus; *Ant*, antenna; *Bk*, beak; *Clp*, clypeus; *e*, roof of mouth cavity,
or sucking pump; *Ge*, gena (cheek plate); *Hphy*, hypopharynx; *Lb*, labium; *Lm*,
labrum; *Md*, base of mandible; *MdB*, mandibular bristle; *Mth*, mouth; *Mx*,
maxilla; *MxB*, maxillary bristle; *NMb*, neck membrane; *O*, ocelli

bristles, small as they are, are fastened together by inter-
locking ridges and grooves, so that their apposed fur-
rows are converted into a single tubular channel. In
the natural position, these second bristles lie in the
sheath of the beak (Fig. 121 A) between the somewhat
larger first bristles. Their bases separate at the tip of

the tongue (*Hphy*) to pass to either side of the latter organ, but the channel between them here becomes continuous with a groove on the middle of the forward surface of the tongue. When the mouth-opening is closed, as it always is in the fully matured insect, the tongue groove is converted into a tube which leads upward from the channel between the second bristles into the inner cavity of the mouth. It is through this minute passage that the cicada obtains its liquid food; but obviously there must be a pumping apparatus to furnish the sucking force.

The sucking mechanism is the mouth cavity and its muscles. The mouth cavity, as seen in a section of the head (Fig. 122, *Pmp*), is a long, oval, thick-walled capsule having its roof, or anterior wall (*e*), ordinarily bent inward so far as almost to fill the cavity. Upon the midline of the roof is inserted a great mass of muscle fibers (*PmpMcls*) that have their other attachment on the striated plate of the face (*Clp*). The contraction of these muscles lifts the roof, and the vacuum thus created in the cavity of the mouth sucks up the liquid food. Then the muscles relax, and the elastic roof again collapses, but the lower end comes down first and forces the liquid upward through the rear exit of the mouth cavity into the pharynx, a small muscular-walled sac (*Phy*) lying in the back of the head. From the pharynx, the food is driven into the tubular gullet, or oesophagus (*OE*), and so on into the stomach.

The bases of both pairs of bristles are retracted into pouches of the lower head wall behind the tongue, and upon each bristle base are inserted sets of protractor and retractor muscle fibers. By means of these muscles, the bristles can be thrust out from the tip of the beak or withdrawn, and the bristles of the stronger first pair are probably the chief organs with which the insect punctures the tissues of the plant on which it feeds. As the bristles enter the wood, the sheath of the beak can be retracted into the flexible membrane of the neck at its base.

One other structure of interest in the cicada's head should be observed. This is a force pump connected with the duct (Fig. 122, *SalD*) of the large salivary glands (*Gl, Gl*) and used probably for injecting into the wound of the plant a secretion which perhaps softens the tissues of the latter as the bristles are inserted. Possibly the saliva has also a digestive action on the food liquid. The salivary pump (*SalPmp*) lies behind the mouth, and its duct opens on the extreme tip of the tongue, where the saliva can be driven into the channel of the second bristles. Most sucking insects have two parallel channels between these bristles (Fig. 90), one for taking food, the other for ejecting saliva, and the cicada probably has two also, though investigators differ as to whether there are two or only one.

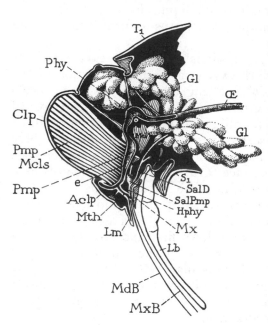

FIG. 122. Median section of the head and beak of an adult cicada

The sucking pump (*Pmp*) is the mouth cavity, the collapsed roof of which (*e*) can be lifted like a piston by the large muscles (*PmpMcls*) arising on the clypeus (*Clp*). The liquid food ascends through a channel between the maxillary bristles (*MxB*), is drawn into the mouth opening (*Mth*), and pumped back into the pharynx (*Phy*), from which it goes into the oesophagus (*OE*). A salivary pump (*SalPmp*) opens at the tip of the hypopharynx (*Hphy*), discharging the secretion of the large glands (*Gl, Gl*) into the beak

The head of the cicada is thus seen to be a wonderful mechanism for enabling the insect to feed on plant sap. The piercing beak and the sucking apparatus, however, are characters distinguishing the members of a whole order of insects, the Hemiptera, or Rhynchota. This order includes, besides the cicadas, such familiar insects as the plant lice, the scale insects, the squash bugs, the giant water bugs, the water striders, and the bed bugs. To the sucking insects properly belongs the name "bug," which is not a synonym of "insect."

It is believed, of course, that the parts of the sucking beak of a hemipteran insect correspond with the mouth parts of a biting insect, described in Chapter IV (Fig. 66), but it has been a difficult matter to determine the identities of the parts in the two cases. Probably the anterior narrow plate on the side of the cicada's head (Fig. 121, *Md*) is a rudiment of the base of the true jaw, or mandible. The first bristles (*MdB*) are outgrowths of the mandibular plates, which have become detached from them and made independently movable by special sets of muscles. The second bristles (*MxB*) are outgrowths of the maxillae, which are otherwise reduced to small lobes (*Mx*) depending from the cheek plates (*Ge*). The sheath of the beak (*Lb*) is the labium. We have here, therefore, a most instructive lesson on the manner in which organs may be made over in form, by the processes of evolution, adapting them to new and often highly special uses.

The abdomen of the cicada is thick, and strongly arched above. Its external appearance of plumpness suggests that it would furnish a juicy meal for a bird, and birds do destroy large numbers of the insects. Yet when the interior of a cicada is examined (Fig. 123), it is found that almost the entire abdomen is occupied by a great air chamber! The soft viscera are packed into narrow spaces about the air chamber, the stomach (*Stom*) being crowded forward into the rear part of the thorax.

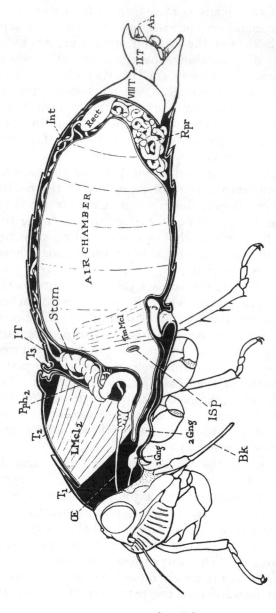

FIG. 123. Vertical lengthwise section through the middle of the body of a male cicada, showing the great air chamber almost filling the abdomen

An, anus; *Bk*, beak; *1Gng*, *2Gng*, first and second nerve centers of the thorax; *Int*, intestine; *IT*, back plate of first abdominal segment; *ISp*, first abdominal spiracle, opening into air chamber; *IXT*, ninth abdominal segment; *j*, plate supporting the tympanal muscles (*TmMcl*); *LMcl₂*, thoracic muscles; *OE*, oesophagus; *Rect*, rectum; *Rpr*, organs of reproduction; *Stom*, stomach; *T₁*, *T₂*, *T₃*, *T₄*, back plates of thorax; *TmMcl*, muscle of right tympanum; *VIIIT*, eighth abdominal segment

The air chamber is a large, thin-walled sac of the tracheal respiratory system, and receives its air supply directly through the spiracles of the first abdominal segment. From the sac are given off tracheal tubes to the muscles of the thorax and to the walls of the stomach.

Many insects have tracheal air sacs of smaller size, and the purpose of the sacs in general appears to be that of holding reserve supplies of air for respiratory purposes. The great size of the air sac in the cicada's abdomen, however, suggests that it has some special function, and it is natural to suppose that it acts as a resonating chamber in connection with the sound-producing drums. Yet the sac is as well developed in the female as in the male. Possibly, therefore, it serves too for giving buoyancy to the insects, for it can readily be seen that if the space occupied by the sac were filled with blood or other tissues, as it is in most other insects, the weight of the cicada would be greatly increased; or, on the other hand, if the body were contracted to such a size as to accommodate only its scanty viscera, it would lose buoyancy through lack of sufficient extent of surface— a paper bag crumpled up drops immediately when released, but the same bag inflated almost floats in the air.

The Sound-Producing Organs and the Song

The cicadas produce their music by instruments quite different from those of any of the singing Orthoptera —the grasshoppers, katydids, and crickets, described in Chapter II. On the body of the male cicada, just back of the base of each hind wing, as we have already observed, in the position of the "ear" of the grasshopper (Fig. 63, *Tm*), there is an oval membrane like the head of a drum set into a solid frame of the body wall (Fig. 120, *Tm*). Each drumhead, or tympanum, is a membrane closely ribbed with stiff vertical thickenings, the number of ribs varying in different species of cicadas and perhaps accounting in part for the different qualities of sound produced. In the periodical cicada, the drum-heads are exposed and are easily seen when the wings are lifted; in our other common cicadas each drum is concealed by a flap of the body wall.

The sound made by an ordinary drum is produced by the vibration of the drumhead that is struck by the player, but the tone and volume of the sound are given by the air space within the drum and by the sympathetic vibration of the opposite head. The air within the drum, then, must be in communication with the air outside the drum, else it would impede the vibration of the drumheads.

All these conditions imposed upon a drum are met by the cicada. The abdomen of the insect, as we have seen, is largely occupied by a great air chamber (Fig. 123), and the air within the chamber communicates with the outside air through the spiracles of the first abdominal segment (*ISp*). In addition to the two drumheads whose activity produces the sound, there are two other thin, taut membranous areas set into oval frames in the lower side walls of the front part of the abdomen (not seen in the figures). These ventral drumheads have such smooth and glistening surfaces that they are often desig-nated the "mirrors." The wall of the air sac is applied closely to their inner surfaces, but both membranes are so thin that it is possible to see through them right into the hollow of the cicada's body. The ventral drum-heads are not exposed externally, however, for they are covered by two large, flat lobes projecting back beneath them from the under part of the thorax.

The cicada does not beat its drums or play upon them with any external part of its body. When a male is "singing," the exposed drumheads are seen to be in very rapid vibration, as if endowed with the power of auto-matic movement. An inspection of the interior of the body of a dead specimen, however, shows that con-nected with the inner face of each drumhead is a thick muscle which arises below from a special support on the ventral wall of the second abdominal segment (Figs. 123, 124, *TmMcl*). It is by the contraction of these muscles that the drum membranes are set in motion.

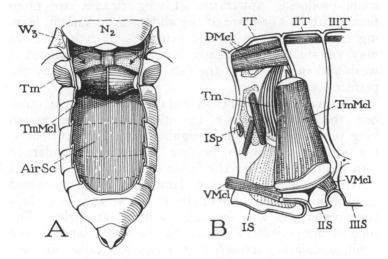

FIG. 124. The abdomen and sound-making organs of the male periodical cicada

A, the abdomen cut open from above, exposing the air chamber (*AirSc*), and showing the great tympanal muscles (*TmMcl*) inserted on the tympana (*Tm*). The arrows indicate the position of the first spiracles opening into the air chamber (see fig. 123, *ISp*)

B, inner view of right half of first and second abdominal segments, showing the ribbed tympanum (*Tm*), and the muscles that vibrate it (*TmMcl*)

AirSc, air chamber; *DMcl*, dorsal muscles; *IS, IIS, IIIS*, sternal plates of first three abdominal segments; *ISp*, first abdominal spiracle; *IT, IIT, IIIT*, tergal plates of first three abdominal segments; *N₂*, tergal plate of third thoracic segment; *Tm*, tympanum; *TmMcl*, tympanal muscle; *W₃*, base of hind wing; *VMcl*, ventral muscles

But a muscle pulls in only one direction; the drum muscles produce directly the inward stroke of the drumhead membranes; the return stroke results from the outward convexity and the elasticity of the heads themselves and the stiff ribs in their walls.

When a cicada starts its music, it lifts the abdomen a little, thus opening the space between its ventral drumheads and the protecting flaps beneath, and the sound comes out in perceptibly increased volume. There can be little doubt that the air chamber of the body and the ventral membranes are important accessories in the sound-producing apparatus. Living cicadas are often found with half or more of the abdomen broken off, leaving the air sac open to the exterior. Such individuals may vibrate the drumheads, but the sound produced is weak and entirely lacks the quality of that made by the perfect insect.

Wherever the periodical cicada appears in great numbers, the daily choruses of the males leave an impression long remembered in the neighborhood; and, curiously, the sound appears to become increasingly louder in retrospect, until, after the lapse of years, each hearer is convinced it was a deafening clamor that almost deprived him of his senses. Fortunately the cicadas are daytime performers and are seldom heard at night. The song of the periodical species has no resemblance to the shrill, undulating screech of the annual cicadas so common every summer in August and September. All the notes of the more common large form of the seventeen-year race are characterized by a *burr* sound, and at least four different utterances may be distinguished; the quality of three of the notes probably depends on the age of the individual insect, the fourth is an expression of fright or anger.

The simplest notes to be heard are soft purring sounds, generally made by solitary insects sitting low in the bushes, probably individuals that have but recently emerged from the ground. The next is a longer and louder note, characterized by a rougher *burr*, lasting about five seconds, and always given a falling inflection at the close. This sound is the one popularly known as the "Pharàoh" song, because of a fancied resemblance to the name if the first syllable is sufficiently prolonged and the second allowed to drop off abruptly at the end. It is repeated at intervals of from two to five seconds, and is given always as a solo by individuals sitting in the bushes or on lower branches of the trees. Males singing the Pharaoh song, therefore, are easily observed in the act of performing. With the beginning of each note, the singer lifts his abdomen to a rigid, horizontal position, thus opening the cavity beneath the lower drumheads and letting out the full volume of the sound. Toward the end of the note, the abdomen drops again to the usual somewhat sagging position, appearing thus to give the abrupt falling inflection at the close.

The grand choruses, by which the periodical cicada is chiefly known and remembered, are given by the fully matured males of the swarm, always high in the trees where the singers may seldom be closely observed while performing. The individual notes are prolonged *bur-r-r-*like sounds, repeated all day and day after day, but all single voices are blended and lost in the continuous hum of the multitude.

The fourth note of the larger form of the cicada is uttered by males when they appear to be surprised or frightened. On such occasions, as the insect darts away, he makes a loud, rough sound, and the same note is often uttered when a male is picked up or otherwise handled.

The notes of the small form of the seventeen-year race of the cicada have an entirely different character from those of his larger relative. The regular song of the little males much more resembles that of the annual summer cicadas, though it is not so long and is less continuous in tone. It opens with a few short chirps; then follows a series of strong, shrill sounds like *zwing, zwing, zwing,* and so on, closing again with a number of chirps. The whole song lasts about fifteen seconds. Several of these males kept in cages for observation sang this song repeatedly and no other. It is common out of doors, but always heard in solo, never in chorus. When handled or otherwise disturbed, the small males utter a succession of sharp chirps very suggestive of the notes of some miniature wren angrily scolding at an intruder. Never does the small form of the cicada utter notes having the *burr* tone of those of the larger species, and the vocal differences of the two varieties are strikingly evident when several males of both kinds are caged together. When disturbed, each produces his own sound, one the burr, the other the chirp; and there is never any suggestion of similarity or of gradation between them.

Egg Laying

The cicadas lay their eggs in the twigs of trees and shrubs and frequently in the stalks of deciduous plants. They show no particular choice of species except that conifers are usually avoided.

The eggs are not stuck into the wood at random, but are carefully placed in skillfully constructed nests which the female excavates in the twigs with the blades of her ovipositor (Plate 8). These nests are perhaps always on the under surfaces of the twigs, unless the latter are vertical, and usually there are rows of from half a dozen to twenty or more of them together.

Egg laying begins in the early part of June, and by the tenth of June it is at its height. The female cicadas can easily be watched at work, taking flight only from actual interference. They usually select twigs of last year's growth, but often use older ones or green ones of the same season. In the majority of cases the female works outward on the twig; but if this is a rule, it is a very loosely observed one, for many work in the opposite direction.

Each nest is double; that is, it consists of two chambers having a common exit, but separated by a thin vertical partition of wood (Plate 8, D, F). The eggs are placed on end in the chambers in two rows, with their head ends

PLATE 8

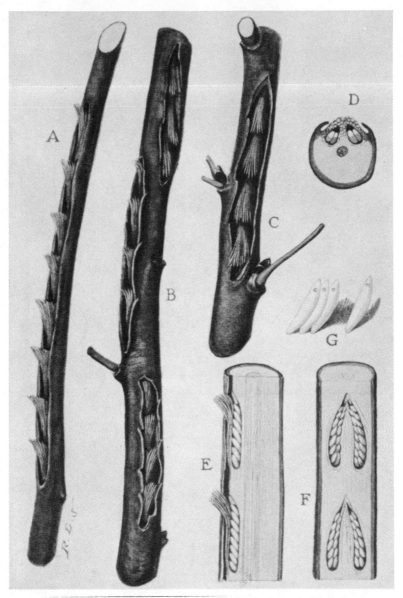

Egg punctures and the eggs of the periodical cicada

A, B, C, twigs of dogwood, oak, and apple containing rows of cicada
egg nests. D, cross-section of a twig through an egg nest, showing the
two chambers, each containing a double row of eggs. E, vertical
lengthwise section through two egg nests, showing the rows of slanting
eggs and the frayed lip of the nest opening. F, horizontal section
showing each chamber filled with a double row of eggs. G, several eggs
(much enlarged)

downward and slanted toward the door. Generally there are six or seven eggs in each row (*E*), making twenty-four to twenty-eight eggs in the whole nest, but frequently there are more than this. The wood fibers at the entrance are much frayed by the action of the ovipositor and make a fan-shaped platform in front of the door (A, B, C). Here the young shed their hatching garments on emerging from the nest. The series of cuts in the bark eventually run together into a continuous slit, the edges of which shrink back so that the row of nests comes to have the appearance of being made in a long groove. This mutilation kills many twigs, especially those of oaks and hickories, the former soon showing the attacks of the insects by the dying leaves. The landscape of oak-covered regions thus becomes spotted all over with red-brown patches which often almost cover individual trees from top to bottom. Other trees are not so much injured directly, but the weakened twigs often break in the wind and then hang down and die.

An ovipositing female (Plate 7) finishes each egg nest in about twenty-five minutes; that is, she digs it out and fills it with eggs in this length of time, for each chamber is filled as it is excavated. A female about to oviposit alights on a twig, moves around to the under surface, and selects a place that suits her. Then, elevating the abdomen, she turns her ovipositor forward out of its sheath and directs its tip perpendicularly against the bark. As the point enters it goes backward, and when in at full length the shaft slants at an angle of about forty-five degrees.

In a number of cases females were frightened away at different stages of their work, and an examination of the unfinished nests showed that each chamber is filled with eggs as soon as it is excavated; that is, the insect completes one chamber first and fills it with eggs, then digs out the other chamber which in turn receives its quota of eggs, and the whole job is done. The female now moves forward a few steps and begins work on another nest, which is completed in the same fashion. Some series consist of only three or four nests, while others contain as many as twenty and a few even more, but perhaps eight to twelve are the usual numbers. When the female has finished what she deems sufficient on one twig, she flies away and is said to make further layings elsewhere, till she has disposed of her 400 to 600 eggs, but the writer made no observations covering this point. Probably the cicada feels it safer not to intrust all her eggs to one tree, on the principle of not putting all your money in the same bank.

Death of the Adults

The din of music in the trees continues with monotonous regularity into the second week of June, by which time the mating season is over. Soon thereafter the performers lose their vitality; large numbers of them drop to the earth where many perish from an internal fungus disease that eats off the terminal rings of the body; others are mutilated and destroyed by birds, and the rest perhaps just die a natural death. Beneath the trees, where a great swarm has but recently given such abundant evidence of life, the ground is now strewn with the dead or dying. A large percentage of the living are in various stages of disfigurement—wings are torn off, abdomens are broken open or gone entirely, mere fragments crawl about, still alive if the head and thorax are intact. In the males often the great muscle columns of the drums are exposed and visibly quivering, and many of the insects, game to the end, even in their dilapidated condition still utter purring remnants of their song.

From now on till the latter part of July, the only evidence of the late swarm of noisy visitors will be the scarred twigs on the trees and bushes that have received the eggs and the red-brown patches of dying leaves that everywhere disfigure the oaks and hickories.

THE PERIODICAL CICADA

The Broods

The two races of the periodical cicada, the seventeen-year and the thirteen-year, together occupy most of the eastern part of the United States, except the northern part of New England, the southeastern corner of Georgia, and the peninsula of Florida. The western limits extend into the eastern part of Nebraska, Kansas, Oklahoma, and Texas. In general, the seventeen-year race is northern, and the thirteen-year race is southern, but, though the geographic line between the two races is remarkably distinct, there is considerable overlapping.

While the two cicada races are distinguished from each other by the length of their life cycle, the members of each race do not all appear in the adult stage in any one year. Both the seventeen-year race and the thirteen-year race are broken up into groups of individuals that emerge in different years, and these groups are known as "broods." Each brood has its definite year of emergence, and in general a pretty well-defined territory. The territories of the different broods, however, overlap, or the range of a small brood may be included in that of a larger one. Hence, in any particular locality, there is not always an interval of thirteen or seventeen years between the appearance of the insects; and it may happen that members of a thirteen-year brood and of a seventeen-year brood will emerge in the same year at the same place.

The emergence years of the principal cicada broods have now been recorded for a long time, and the oldest record of a swarm is that of the appearance of the "locusts" in New England two hundred and ninety-five years ago. A full account of the broods of both races of the periodical cicada, their distribution, and the dates of their emergence, is given in Dr. C. L. Marlatt's Bulletin, already cited, and the following abstract is taken from this source:

Wherever a well-defined cicada brood appears in a certain year, it is generally observed that a few individuals

[215]

come out the year before or the year after. This fact has suggested the idea that the various broods established at the present time had their origin from individuals of a primary brood that, as we might say, got their dates mixed, and came out a year too soon or a year too late, the multiplying descendants of these individuals thus founding a new brood dated a year in advance or a year behind the emergence time of the parent stock. In this way, it is conceivable, the seventeen-year race might come to appear on each of seventeen consecutive years, and the thirteen-year race on each of thirteen consecutive years. Individuals emerging on the eighteenth or fourteenth year, according to the race, would be reckoned as a part of the first brood of its race.

The facts known concerning the emergence of the cicadas seem to confirm the above theory, for members of the seventeen-year race appear somewhere every year within the limits of their range, and the emergence of members of the thirteen-year race has been recorded for at least eleven out of the possible thirteen years. All the individuals of a brood are not, of course, descendants of a single group of ancestors, nor do they necessarily occur together in a restricted area—they are simply individuals that coincide in the year of their emergence. However, at least thirteen of the broods of the seventeen-year race are well defined groups, for the most part with definitely circumscribed territories, though overlapping in many cases. The broods of the thirteen-year race are not so well developed.

The broods are conveniently designated by Roman numerals. According to the system of brood numbering proposed by Doctor Marlatt, and now generally adopted, the brood of the seventeen-year race that appeared last in 1927 is Brood I. This is not a large brood, but it has representatives in Pennsylvania, Maryland, District of Columbia, Virginia, West Virginia, North Carolina, Kentucky, Indiana, Illinois, and eastern Kansas. Brood II, 1928, lives in the Middle Atlantic States, with a few scattering colonies farther west. Brood III, 1929, is mostly confined to Iowa, Illinois, and Missouri. The largest of the broods is X, covering almost the entire range of the seventeen-year race. This brood made its last appearance in 1919, and is due next, therefore, in 1936. The series of broods as numbered thus follows the successive years to Brood XVII, the last brood of the seventeen-year race, which will return next in 1943.

The small and uncertain broods of the seventeen-year race are VII, XII, XV, XVI, and XVII. The cicadas that emerge in the years corresponding with these numbers represent incipient broods, being probably the descendants of a few individuals that sometime became separated from the larger broods of the years preceding or following. One of the smallest of the seventeen-year broods is XI, but since its colonies occur in Massachusetts, Connecticut, and Rhode Island, it is likely that it was more numerous in individuals in former times than at present. The brood with the oldest recorded history is XIV. This is a large brood extending over much of the range of the seventeen-year race, with colonies in eastern Massachusetts on Cape Cod and near Plymouth, the emergence of which was observed by the early settlers probably in 1634.

The broods of the thirteen-year race are numbered from XVIII to XXX, Brood XVIII being that which appeared last in 1919. But there are only two important broods of this southern race, XIX, which emerged in 1920, and XXII, which emerged in 1924. In most of the other years the shorter-lived race is represented by only a few individuals that emerge here and there over its range; and none at all are known to appear during the years corresponding with the numbers XXV and XXVIII.

THE HATCHING OF THE EGGS

Five weeks have elapsed since the departure of the cicada swarms. It is nearly six weeks since egg laying was at its height, and the eggs are now due to hatch almost any time. When studying the cicadas of Brood X near Washington in 1919, the writer found the first evidence of hatching on the twenty-fourth of July. Perhaps the normal time of hatching had been delayed somewhat by heavy rains that fell almost continuously during the ten days previous, for many eggs examined during this time were found to be dead and turning brown, though the percentage of these was small. The twenty-fifth was hot and bright all day. The trees were inspected in the afternoon. Their twigs had been bare the day before. Now, at the entrance holes of the egg nests were little heaps of shriveled skins, thousands in all, and each so light that the merest breath of air sufficed to blow it off; so, if according to this evidence thousands of nymphs had hatched and gone, the evidence of as many more must have been carried away by the winds. An examination of many egg nests themselves showed that over half contained nothing but empty shells. Whole series were thus deserted, and usually all or nearly all, of the eggs in any one series of nests would be either hatched or unhatched. But often the eggs of one or more nests would be unhatched or mostly so in a series containing otherwise only empty shells. Delay appeared to go by nests rather than by individual eggs.

As a very general rule the eggs nearest the door of an egg chamber are the ones that hatch first, the others following in succession, though not in absolute order. But unhatched eggs, if present, are always found at the bottom of the nest, with the usual exception of one or two farther forward. Only occasionally an empty shell occurs in the middle of an unhatched row. If the actual hatching of the eggs is observed in an opened nest, several nymphs are usually seen coming out at the same time, and in nearly all cases they are in neighboring eggs, though not always contiguous ones. So this rule of hatching, like most rules, is general but not binding.

The procedure of the female in placing the eggs leaves no doubt that the first-laid ones are those at the bottom of the cell, showing that the order of laying has no relation to the order of hatching, except that it is mostly the reverse. It seems hardly reasonable to suppose that the eggs nearest the door are affected by greater heat or by a fresher supply of air, so it is suggested that the order of hatching may be due simply to the successive release of pressure along the tightly packed rows, giving the compressed embryos a chance to squirm and kick enough to split the inclosing shells. When hatching once commences it proceeds very rapidly through the whole nest, showing that the eggs are all at the bursting point when the rupture of the first takes place.

In each lateral compartment of an egg nest the eggs (Plate 8, E, F) stand in two rows with their lower or head ends slanted toward the door. (It must be remembered that the punctures are made on the lower sides of the twigs, so that the eggs are inverted in their natural position in the nests.) On hatching, each egg splits vertically over the head and about one-third of the length along the back, but for only a short distance on the ventral side. As soon as this rupture occurs, the head of the young cicada bulges out; and then, by a bending of the body back and forth, the creature slowly works its way out of the shell, which, when empty, remains behind in its original place. The nymphs nearest the door have an easy exit, but those from the depths of the cell find themselves still in a confined space between the projecting ends of the empty shells ahead of them and the chamber wall, a passage almost as narrow as the egg itself, through which the delicate creatures must squirm to freedom.

A newly-hatched or a newly-born aphid, as we have seen in Chapter VI, is done up in a tight-fitting garment with neither sleeves nor legs, but nature has been more considerate in the case of the young cicada. It, too, comes out of the egg clothed in a skin-tight jacket, but

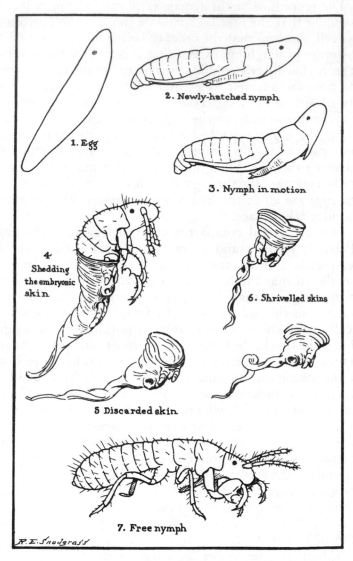

1. Egg

2. Newly-hatched nymph

3. Nymph in motion

4. Shedding the embryonic skin

5 Discarded skin

6. Shrivelled skins

7. Free nymph

R. E. Snodgrass

FIG. 125. The egg, the newly-hatched nymph shedding the embryonic
skin, and the free nymph of the periodical cicada

this garment is not a mere bag; it is provided with special pouches for the appendages or a part of them (Fig. 125, 2). The incased antennae and the labrum project backward as three small points lying against the breast. The front legs are free to the bases of the femora, though so tightly held in their narrow sleeves that their joints have no independent motion. The middle and hind legs are also incased in long, slim sheaths, but they always adhere close to the sides of the body. Thus the cicada nymph newly-hatched much resembles a tiny fish provided only with two sets of ventral fins, but when it gets into action its motions are comparable with the clumsy flopping of a seal stranded on the beach and trying to get back into the water (3).

The infant cicada knows it is not destined to spend its life in the narrow cavern of its birth, or at least it has no desire to do so. With its head pointed toward the exit, it begins at once contortionistic bendings of the body, which slowly drive it forward. By throwing the head and thorax back, the antennal tips and the front legs are made to project so that their points may take hold on any irregularity in the path. Then a contractile wave running forward through the abdomen brings up the rear parts of the body as the front parts are again bent back, and the "flippers" grasp a new point of support. As these motions are repeated over and over again, the tiny, awkward thing painfully but surely moves forward, perhaps helped in its progress by the inclined tips of the flexible eggshells pressing against it, on the same principle that a head of barley automatically crawls up the inside of your sleeve.

Once out of the door no time is lost in discarding the encumbering garment, but it is never shed in the nest under normal conditions. If, however, the nest is cut open and the hatching nymph finds itself in a free, open space, the embryonic sheath is cast off immediately, often while the posterior end of the insect's body is still in the egg, so that the skin may be left sticking in the open end of the shell. If the young cicada did not have to gain its liberty through that narrow corridor, it might be born in a smooth bag as are its relations, the aphids.

Watching at the door of an undisturbed nest during a hatching day, we soon may see a tiny pointed head come poking out of the narrow hole. The threshold is soon crossed, but no more; this traveling in a bag is not a pleasure trip. A few contortions are always necessary to rupture the skin, and sometimes several minutes are consumed in violent twistings and bendings before it splits. When it does break, a vertical rent is formed over the top of the head, which latter bulges out until the cleft becomes a circle that enlarges as the entire head pushes through, followed rapidly by the body (Fig. 125, 4). The appendages come out of their sheaths like fingers out of a glove, turning the pouches outside in. The antennae are free first; they pop out and hang stiffly downward. Then the front legs are released and hang stiff and rigid but quivering with a violent trembling. In a second or so this has passed, the joints double up and assume the characteristic attitude, while they violently claw the air. Then the other legs and the abdomen come out and the embryo is a free young cicada (7). All this usually happens in less than a minute, and the new creature is already off without so much as a backward glance at the clothes it has just removed or at the home of its incubation period. Sentiment has no place in the insect mind.

As the nymphs emerge from the nest, one after another, and shed their skins, the glistening white membranes accumulate in a loose pile before the entrance, where they remain until wafted off on the breeze. Each discarded sheath has a goblet form (Fig. 125, 5, 6), the upper stiff part remaining open like a bowl, the lower part shriveling to a twisted stalk. The antennal and labral pouches project from the skin as distinct appendages, but those of the legs are usually inverted during the shedding and disappear from the outside of the slough, though the holes where they were pulled in can be found before the membrane becomes too dry.

The nymph (Figs. 125, 7; 126) usually runs about at first in the groove of the twig containing its egg nest and then goes out on the smooth bark. Here any current of air is likely to carry it off immediately, but many wander about for some time, usually going toward the tips of the twigs, some even getting clear out on the leaves. But only a few nymphs are ever to be found on twigs where piles of embryonic skins show that hundreds have recently hatched; so it is evident that the great majority either fall off or are blown away very shortly after emerging. Many undoubtedly fall before the shedding of the egg membrane, for the inclosed creature has no possible way of holding on, and even the free nymph has but feeble clinging powers. Those observed on twigs kept indoors often fell helplessly from the smooth bark while apparently making real efforts to retain their grasp. Their weak claws could get no grip on the hard surface. Instead, then, of deliberately launching themselves into space in response to some mysterious call from below, the young cicadas simply fall from their birthplace by mere inability to hold on. But the same end is gained— they reach the ground, which is all that matters. Nature is ever careless of the means, so long as the object is attained. Some acts of unreasoning creatures are assured by bestowing an instinct, others are forced by withholding the means of acting otherwise.

The cicada nymphs are at first attracted by the light. Those allowed to hatch on a table in a room will leave the twigs and head straight for the windows ten feet away. This instinct under natural conditions serves to entice the young insects toward the outer parts of the tree, where they have the best chance of a clear drop to earth; but even so, adverse breezes, irregularity of the trees, underbrush, and weeds can not but make their downward journey one of many a bump and slide from leaf to leaf before the earth receives them.

The creatures are too small to be followed with the eye as they drop, and so their actual course and their behavior when the ground is reached are not recorded. But several hatched indoors were placed on loose earth packed

FIG. 126. The young cicada nymph ready to enter the ground (greatly enlarged)

flat in a small dish. These at once proceeded to get below the surface. They did not dig in, but simply entered the first crevice that they met in running about. If the first happened to terminate abruptly, the nymph came out again and tried another. In a few minutes all had found satisfactory retreats and remained below. The eagerness with which the insects dived into any opening that presents itself indicates that the call to enter the earth is instinctive and imperative once their feet have touched the ground. Note, then, how within a few minutes their instincts shift to opposites: on hatching, their first effort is to extricate themselves from the narrow confines of the egg nest, and it seems unlikely that enough light can penetrate the depths of this chamber to guide them to the exit; but once out and divested of their encumbering embryonic clothes, the insects are irresistibly drawn in the direction of the strongest light, even though this takes them upward—just the opposite of their destined course. When this instinct has served its purpose and has taken the creatures to the port of freest passage to the earth, all their love of light is lost or swallowed up in the call to enter some dark crevice narrower even than the one so recently left by such physical exertion.

When the young cicadas have entered the earth we practically have to say good-bye to them until their return. Yet this recurring event is ever full of interest to us, for, much as the cicadas have been studied, it seems that there is still plenty to be learned from them each time they make their visit to our part of the world.

CHAPTER VIII

INSECT METAMORPHOSIS

THE fascination of mythology and the charm of fairy tales lie in the power of the characters to change their form or to be changed by others. Zeus would court the lovely Semele, but knowing well she could not endure the radiance of a god, he takes the form of a mortal. Omit the metamorphosis, and what becomes of the myth? And who would remember the story of Cinderella if the fairy godmother were left out? The flirtation between the heroine and the prince, the triumph of beauty, the chagrin of the haughty sisters—these are but ingredients in the pot of common fiction. But the transformation of rats into prancing horses, of lizards into coachman and lackeys, of rags into fine raiment—this imparts the thrill that endures a lifetime!

It is not surprising, then, that the insects, by reason of the never-ending marvel of their transformations, hold first place in every course of nature study in our modern schools, or that nature writers of all times have found a principal source of inspiration in the "wonders of insect life." Nor, finally, should it be made a matter of scorn if the insects have attached themselves to our emotions, knowing how ardently the natural human mind craves a sign of the supernatural. The butterfly, spirit of the lowly caterpillar, has thus been exalted as a symbol of human resurrection, and its image, carved on graveyard gates, still offers hope to those unfortunates interred behind the walls.

Metamorphosis is a magic word, in spite of its formidable

[226]

appearance; but rendered into English it means simply "change of form." Not every change of form, however, is a metamorphosis. The change of a kitten into a cat, of a child into a grown-up, of a small fish into a large fish are not examples of metamorphosis, at least not of what is called metamorphosis. There must be something spectacular or unexpected about the change, as in the transformation of the tadpole into a frog, the change of the wormlike caterpillar into a moth, or of a maggot into a

FIG. 127.   Moths of the fall webworm

fly. This arbitrary limiting of the use of a word that might, from its derivation, have a much more general meaning, is a common practice in science, and for this reason every scientific term must be defined. Metamorphosis, then, as it is used in biology, signifies not merely a change of form, but a particular kind or degree of change; the kind of change, we might say, that would appear to lie outside the direct line of development from the egg to the adult.

At once it becomes evident that, by reason of the very definition we have adopted, our subject is going to become complicated; for how are we to decide if an observed change during the growth of an animal is in line or out of line with direct development? There, indeed, lies a serious difficulty, and we can only leave it to the biologist to decide in any particularly doubtful case. But there are plenty of cases concerning which there is no doubt. A caterpillar, for example, certainly is not a form headed toward a butterfly in its growth, and yet we know it is a young butterfly because it hatches out of the butterfly's egg. And, as the caterpillar grows from a small caterpillar to a large caterpillar, it becomes no more like a butterfly than it was at first. It is only after it has reached maturity as a caterpillar that it undergoes a process of transformation by which it attains at last the form of the insect that produced it.

The question now arises as to whether the butterfly is a form superadded to the caterpillar, or the caterpillar a form that has deviated from the developmental line of its ancestors. This question is easily answered: the butterfly represents the true adult form of its species, for it has the essential structure of all other insects, and it alone matures the sexual organs and acquires the power of reproduction. The caterpillar is an aberrant form that somehow has been interpolated between the egg and the adult of its kind. The real metamorphosis in the life of the butterfly, therefore, is not the change of the caterpillar into the adult, but the change of the butterfly embryo in the egg into a caterpillar. Yet the term is usually applied to the reverse process by which the caterpillar is turned back into the normal form of its species.

The caterpillar and the butterfly (Fig. 128) furnish the classical example of insect metamorphosis. Many other insects, however, undergo the same kind of transformation. All the moths as well as the butterflies are caterpillars when they are young: the famous giant moths (Plate 10), including the *Cecropia*, the *Promethea*, and the beautiful *Luna* (Fig. 129), as every nature student knows, come from huge fat caterpillars; the humble cutworms (Fig. 130), when their work of destruction is completed, change into those familiar brown or gray furry moths of moderate size (A) often found hidden away in the daytime and attracted to lights at night. In the spring, the

PLATE 9

Two species of large moths, natural size, showing the beautiful markings with which even night-flying insects may be adorned. Upper figure, *Heliconisa arpi* Schaus, from Brazil; lower, *Dirphia carminata* Schaus, from Mexico. (From J. M. Aldrich)

Fig. 128. The cellery caterpillar, and the butterfly into which it transforms

May beetles, or "June bugs" appear (Fig. 131 A); they are the parents of the common white grubs (B) which every gardener will recognize. The common ladybird beetles (Fig. 132 A) are the adults of the ugly larvae (D) that feed so voraciously on aphids. In the comb of the beehive or

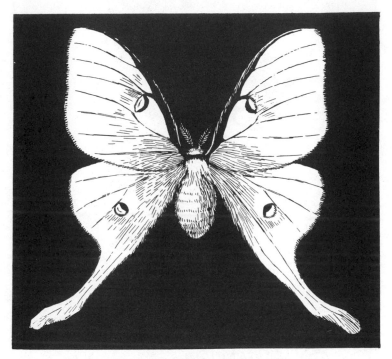

FIG. 129. The Luna moth

of the wasps' nest, there are many cells that contain small, legless, wormlike creatures; these are the young bees or wasps, but you would never know it from their structure, for they have scarcely anything in common with their parents (Fig. 133 A, B). The young mosquito (Fig. 174 D) we all know, from seeing it often pictured and described and from observing that mosquitoes abound wherever these wigglers are allowed to live. The young

PLATE 10

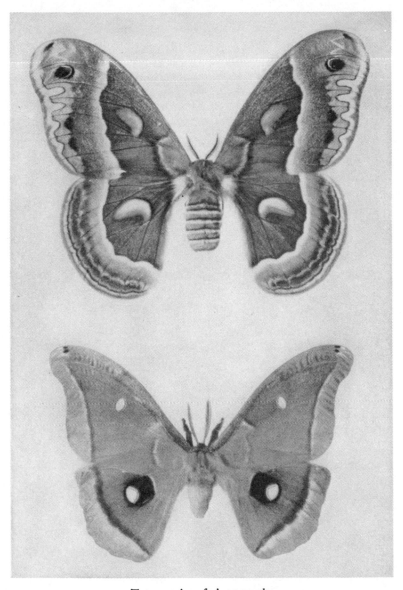

Two species of giant moths
Upper figure, the cecropia moth, female; lower, the polyphemus moth,
male. (From A. H. Clark)

fly is a maggot (Fig. 182 D). The maggots of the house fly inhabit manure piles; those of the blow fly live in dead animals where they feed on the decaying flesh.

We might go on and fill a whole chapter, or a whole book for that matter, with descriptions of the forms that insects go through in their metamorphoses, but since other writers have demonstrated that this can be done and without ex-

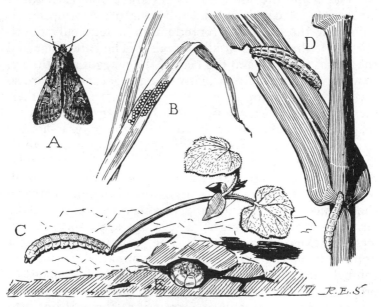

FIG. 130.   The life of a cutworm

A, the parent moth.   B, eggs laid by the moth on a blade of grass.   C, a cutworm at its characteristic night work, eating off a young garden plant at the root.   D, other cutworms climbing the stalk of plants to feed on the leaves. E, the cutworm hidden within the earth during the day

hausting the subject, we shall rather turn our attention here to what may be regarded as the deeper and more abstruse phases of insect metamorphosis.   Where the facts themselves are highly interesting, the explanation of them must be still more so.   Explanations, however, are always more difficult to present than the facts that are to be explained, and if a writer often does not succeed so well with the reader in this undertaking, the reader should remember that his own difficulties of reading are perhaps no greater than the difficulties of the writer in writing. With a little extra effort on both sides, then, we may be able to arrive at a mutual understanding.

In the first place, let us see in what particular manner the young and the adults of insects differ from each other. The adult, of course, is the fully matured form, and it alone has the organs of reproduction functionally developed; but this is true of all animals. The caterpillar and the moth, the grub and the beetle, the maggot and the fly, however, differ widely in many other respects, and are so diverse in appearance and in general structure that their identities can be known only by observing their transformations. On the other hand, the young grasshopper (Fig. 8), the young roach (Fig. 51), or the young aphis (Fig. 97) is so much like its parents that its family relationships are apparent on sight. Still, in the case of all winged insects, there is one persistent difference between the young and the adult, and this is with respect to the development of the wings. The wings are always imperfect or lacking in the young. The inability to fly puts a limitation on the activities of the immature insect and compels it to seek its living by more ordinary modes of progression. It may inhabit the land or the water; it may live on the surface; it may burrow into the earth or into the stems or wood of plants—in short, it may live in a thousand different places, wherever legs or squirming movements will take it, but it can not invade the air, except as it may be carried by the wind.

As a first principle in the study of metamorphosis, then, we must recognize the fact that *only the adult insect is capable of flight.*

Let us now turn back to the grasshopper (Chapter I); it furnishes a good example of an insect in which the adults differ but little from the young, except in the matter of the wings and the organs of reproduction. As might be expected, therefore, the young grasshoppers and the adults live in the same places and eat the same kinds of food in the same way. This likewise is true of the roaches, the katydids, the crickets, the aphids, and other related

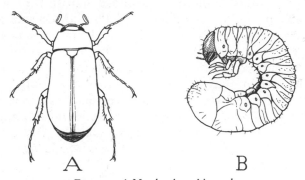

FIG. 131. A May beetle and its grub
A, the adult beetle which feeds on the leaves of shrubs and trees.
B, the larva, a white grub, which lives in the ground and feeds
on roots

insects. The adults here take no advantage over the young in matters of everyday life by reason of their wings.

In many other insects, however, the adults have adopted new ways of living and particularly of feeding, made possible and advantageous to them because of their power of flight. Then, in adaptation to their new habits, they have acquired a special form of the body, of the mouth parts, or of the alimentary canal. But all such modifications, if thrust upon the young, would only be an impediment to them, because the young are not capable of flight. Take the dragonflies as an example. The adult dragonfly (Fig. 58) feeds on small insects which it catches in the air, and it can do so because it has a powerful flying mechanism. The young dragonfly (Figs. 59, 134), however, could not follow the feeding habits of its parents; if it had to inherit the parental form of body and mouth parts

[233]

it would be greatly handicapped for living its own life, and this would be quite as detrimental to the adult, which must be developed from the young. Therefore, nature has devised a scheme for separating the young from the adult, by which the latter is allowed to take full advantage of its wings without imposing a hardship or a disability on its flightless offspring. The device sets aside the ordinary workings of heredity and makes it possible for a structural modification to be developed in the adult and to be suppressed in the young until the time of change from the last immature stage to that of the adult.

Thus we may state as a second principle of metamorphosis that *an adult insect may develop structural characters adaptive to habits that depend on the power of flight, which are suppressed in the young, where they would be detrimental by reason of the lack of wings.*

When parents, now, assert their independence, what can we expect of the offspring? Certainly only a similar declaration of rights. A young insect, once freed from any obligation to follow in the anatomical footsteps of its progenitors, so long as it finally reverts to the form of the latter, soon adopts habits of its own; and then acquires a form, physical characters, and instincts adapted to such habits. Thus, the young dragonfly (Fig. 134) has departed from the path of its ancestors; it has adopted a life in the water, where it feeds upon living creatures which it pursues by its perfection in the art of swimming and captures by a special grasping organ developed from the under lip (B). Life in the water, too, entails an adaptation for aquatic respiration. All the special acquisitions in the structure of the young insect, however, must be discarded at the time of its change to the adult.

A third principle, then, which follows somewhat as a corollary from the second, shows us that *the young of insects may adopt habits advantageous to themselves, and take on adaptive structures that have no regard to the form of the adult and that are discarded at the final transformation.*

The degree of departure of the young from the parental form varies much in different insects. In the cicada, for example, the nymph is not essentially different in structure from the adult except in the matter of the wings, the organs of reproduction and egg laying, and the musical

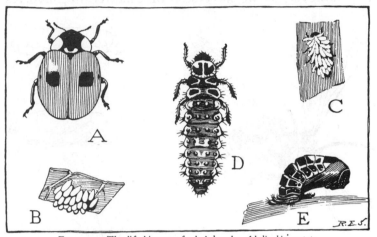

Fig. 132. The life history of a ladybeetle, *Adalia bipunctata*
A, the adult beetle. B, group of eggs on under surface of a leaf. C, a young larval beetle covered with white wax. D, the full-grown larva. E, the pupa attached to a leaf by the discarded larval skin

instrument. But the habitats of the two forms are widely separated, and it is unquestionable that, in the case of the cicada, it is the nymph that has made the innovation in adopting an underground life, for with most of the relatives of the cicada the young live practically the same life as the adults.

Animals live for business, not for pleasure; and all their instincts and their useful structures are developed for practical purposes. Therefore, where the young and the adult of any species differ in form or structure, we may be sure that each is modified for some particular purpose of its own. The two principal functions of any animal are the obtaining of food for its own sustenance, and the production of offspring. The adult insect is necessarily the reproductive stage, but in most cases it must support itself as well; the immature insect has no other direct object in life than that of feeding and of preparing itself for its transformation into the adult. The feeding function, however, as we have seen in Chapter IV, involves

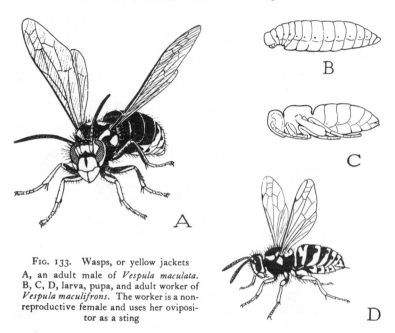

FIG. 133. Wasps, or yellow jackets
A, an adult male of *Vespula maculata*. B, C, D, larva, pupa, and adult worker of *Vespula maculifrons*. The worker is a non-reproductive female and uses her ovipositor as a sting

most of the activities and structures of the animal, including its adaptation to its environment, its modes of locomotion, its devices for avoiding enemies, its means of obtaining food. Hence, in studying any young insect, we must understand that we are dealing almost exclusively with characters that are adaptive to the feeding function.

When we observe the life of any caterpillar we soon realize that its principal business is that of eating. The caterpillar is one creature, at least, that may openly proclaim it lives to eat. Whatever else it does, except acts connected with its transformation, is subservient to the function of procuring food. Most species feed on plants and live in the open (Fig. 135 A); but some tunnel into the leaves (B), into the fruit (D), or into the stem or wood (C). Other species feed on seeds, stored grain, and cereal preparations. The caterpillars of the clothes moths, however, feed on animal wool, and a few other caterpillars are carnivorous.

The whole structure of the caterpillar (Fig. 136) betokens its gluttonous habits. Its short legs (*L*, *AbL*) keep it in close contact with the food material; its long, thick, wormlike body accommodates an ample food storage and gives space for a large stomach for digestive purposes; its hard-walled head supports a pair of strong jaws (*Md*), and since the caterpillar has small use for eyes or antennae, these organs are but little developed. The muscle system of the caterpillar presents a wonderful exhibition of complexity in anatomical structure, and gives the soft body of the insect the power of turning and twisting in every conceivable manner. In contrast to the caterpillar, the moth or the butterfly feeds but little, and its food consists of liquids, mostly the nectar of flowers, which is rich in sugars and high in energy-giving properties but contains little or none of the tissue-building proteins.

When we examine the young of other insects that differ markedly from the parent form, we discover the same thing about them, namely, the general adaptation of their body form and of their habits to the function of eating. Not all, however, differ as widely from the parent as does the caterpillar from the moth. The young of some beetles, for example (Fig. 137), more closely resemble the adults except for the lack of wings. Most of the adult beetles, too, are voracious feeders, and are perhaps not outdone in food consumption by the young. But here another advantage of the double life is demonstrated, for usually the grub and the adult beetle have different modes of life and live in quite different kinds of places. Each individual of the species, therefore, occupies at different times two distinct environments during its life and derives advantages from each. It is true that with some beetles, the young and the adults live together.

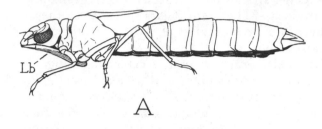

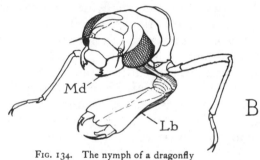

FIG. 134. The nymph of a dragonfly

A, the entire insect, showing the long underlip, or labium (*Lb*), closed against the under surface of the head. B, the head and first segment of the thorax of the nymph, with the labium ready for action, showing the strong grasping hooks with which the insect captures living prey

Such cases, however, are only examples of the general rule that all things in nature show gradations; but this condition, instead of upsetting our generalizations, furnishes the key to evolution, by which so many riddles may be solved.

The grub of the bee or the wasp (Fig. 133 B) gives an excellent example of the extreme specialization in form that the young of an insect may take on. The creature spends its whole life in a cell of the comb or the nest where

[238]

it is provided with food by the parents. Some of the wasps store paralyzed insects in the cells of the nest for the young to feed on; the bees give their young a diet of honey and pollen, with an admixture of a secretion from a pair of glands in their own bodies. The grubs have nothing to do but to eat; they have no legs, eyes, or antennae; each is a mere body with a mouth and a stomach. The adult bees consume much honey, which, like its constituent, nectar, is an energy-forming food; but they also eat a considerable quantity of protein-containing pollen. Yet it is a great advantage to the bees in their social life to have their young in the form of helpless grubs that must stay in their cells until full-grown, when, by a quick transformation, they can take on the adult form and become at once responsible members of the community. Any parents distracted by the incorrigibilities of their offspring in the adolescent stage can appreciate this.

The young mosquito (Fig. 174 D, E) lives in the water, where it obtains its food, which consists of minute particles of organic matter. Some species feed at the surface, others under the surface or at the bottom of the water. The young mosquito is legless and its only means of progression through the water is by a wiggling movement of the soft cylindrical body. It spends much of its time, however, just beneath the surface, from which it hangs suspended by a tube that projects from near the rear end of the body. The tip of the tube just barely emerges above the water surface, where a circlet of small flaps spread out flat from its margin serves to keep the creature afloat. But the tube is primarily a respiratory device, for the two principal trunks of the tracheal system open at its end and thus allow the insect to breathe while its body is submerged.

The adult mosquito (Fig. 174 A), as everybody knows, is a winged insect, the females of which feed on the blood of animals and must go after their victims by use of their wings. It is clear, therefore, that it would be quite impossible for a young mosquito, deprived of the power of flight, to live the life of its parents and to feed after the manner of its mother. Hence, the young mosquito has adopted its own way of living and of feeding, and this has allowed the adult mosquitoes to perfect their specialties without inflicting a hereditary handicap on their offspring. Thus again we see the great advantage which the species as a whole derives from the double life of its individuals.

The fly will only give another example of the same thing. The specialized form of the young fly, the maggot (Fig. 171), which is adapted to the requirements of quite a different kind of life from that of the adult fly, relieves the latter from all responsibility to its offspring. As a consequence, the adult fly has been able to adapt its structure, during the course of evolution, to a way of living best suited to its own purposes, unhampered as it would be if its characters were to be inherited by the young, to whom they would become a great impediment, and probably a fatal handicap.

A fourth principle of metamorphosis, then, we may say, is that *the species as a whole has acquired an advantage by a double mode of existence, which allows it to take advantage of two environments during its lifetime, one suited to the functions of the young, the other to the functions of the adult.*

We noted, in passing, that the young insect is free to live its own life and to develop structures suited to its own purposes under one proviso, which is that it must eventually revert to the form of the adult of its species. At the period of transformation, the particular characters of the young must be discarded, and those of the adult must be developed.

Insects such as the grasshoppers, the katydids, the roaches, the dragonflies, the aphids, and the cicadas appear in the adult form when the young sheds its skin for the last time. The change that has produced the adult, however, began at an earlier period, and the apparently new creature was partially or almost entirely formed within the old skin before the latter was finally shed.

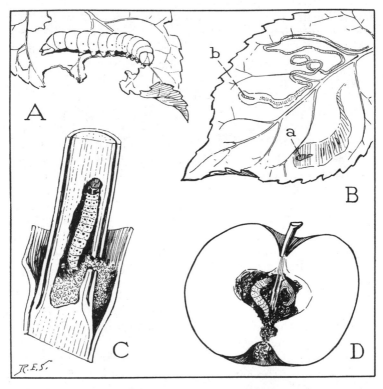

FIG. 135. Various habitats of plant-feeding caterpillars

A, a caterpillar feeding in the open on a leaf. B, leaf miners in an apple leaf, the trumpet miner at *a*, the serpentine miner at *b*. C, the corn borer feeding within a corn stalk. D, the apple worm, or larva of the codling moth, feeding at the core of an apple

After the molt, only a few last alterations in structure and some final adjustments are made while the wings and legs of the creature that had been confined in the closely fitting skin expand to their full length. The structural changes accomplished after the molt, however, vary with different species of insects, and with some they involve a considerable degree of actual growth and change in the form of certain parts. The true transformation process, then, is really a period of rapid reconstructive growth preceding and following the molt, in which the shedding of the skin is a mere incident like the raising of the curtain for a new act in a play. During the intermission the actors have changed their costumes, the old scenery has been removed, and the new has been set in place. Thus it is

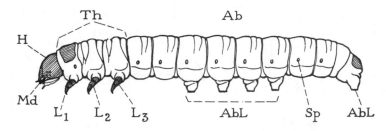

Fig. 136. External structure of a caterpillar
Ab, abdomen; *AbL*, abdominal legs; *H*, head; *L₁, L₂, L₃*, the thoracic legs; *Md*, jaws; *Sp*, breathing apertures; *Th*, thoracic segments

with the insect at the time of its transformation—the special accouterments of the young have been removed, and those of the adult have been put on.

The life of the insect, however, would not make a good theatrical production; it is too much of the nature of two plays given by the same set of actors. The young insect is dressed for a performance of its own in a stage setting appropriate to its act; the adult gives another play and is costumed accordingly. The actor is the same in each case only in the continuity of his individuality. His rehabilitation between the two acts will differ in degree according to the disparity between the parts he plays, that is, according to how far each impersonation is removed from his natural self.

It is evident, therefore, that *the transformation changes of an insect will differ in degree, or quantity, according to the sum of the departure of the young and the departure of the adult from what would have been the normal line of development if neither had become structurally adapted to a special kind of life.*

We may express this idea graphically by a diagram (Fig. 138), in which the line *nm* represents what might have been the straight course of evolution if neither the adult (*I*) nor the young (*L*) had departed along special lines of their own.   But, when the adult and the young have diverged from some point (*a*) in their past history, the line *LI*, which is the sum of *nm* to *L* and of *nm* to *I*, represents the change which the young is bound to make in reverting to the adult form.   The young must, therefore, prepare itself for this event in proportion as the distance *LI* is short or long.

Where the structural disparity between the young and the adult is not great, or is mostly in the external form of the body, the young insect changes directly into the adult, as we have seen in the case of the grasshopper (Fig. 9) and the cicada (Fig. 118).   But with many insects, either because of the degree of difference that has arisen between the young and the adult, or for some other reason, the processes of transformation are not accomplished so quickly and require a longer period for their completion.   In such cases, the creature that issues at the last shedding of the skin by the young insect is in a very unfinished state, and must yet undergo a great amount of reconstruction before it will attain the form and structure of the fully adult insect.   This happens in all the groups of the more highly evolved insects, including the beetles; the moths and butterflies; the mosquitoes and flies; the wasps, bees, ants; and others.   The newly transformed insect must remain in a helpless condition without the use of its legs and wings for a period of time varying in length with different species, until the adult organs, particularly the muscles, are completely formed.

In the meantime, however, the soft cuticular layer of

the skin of the newly emerged insect has hardened, thus preventing a further growth or change in the cellular layer of the body wall beneath it. Reorganization can proceed within the body, but the outer form is fixed and

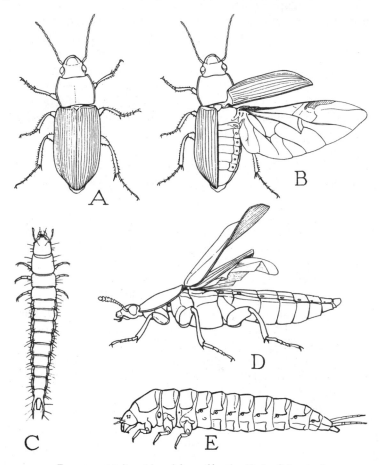

FIG. 137. Adult and larval form of beetles (Order Coleoptera)

A, a ground beetle, *Pterosticus*. B, the same beetle with the right wings spread. C, the larva of *Pterosticus*. D, an adult beetle, *Silpha surinamensis*, with the left wings elevated. E, the larva of the same species, showing the similarity in structure to the adult (D) except for the lack of wings and the shortness of the legs

must remain at the stage it had reached when the cuticula hardened. Only by a subsequent separation of this cuticula, allowing another period of growth in the cells of the body wall, can the form and the external organs of the adult be perfected. With another molt, therefore, the fully formed insect is at last set free, and it now requires only a short time for the expansion of the legs and wings to their normal size and shape and for the hardening of the final cuticular layer which will preserve the contours of the adult.

It thus comes about that *the members of a large group of insects have acquired an extra stage in their life cycle, namely, a final reconstructive stage beginning some time before the last molt of the young and completed with a final added molt which liberates the fully formed adult.* The insect in this stage is called a *pupa.* The entire pupal stage is divided by the last molting of the young into a propupal period, still occupying the loosened cuticula of the insect in its last adolescent stage, and a true pupal period, which is that between the shedding of this last skin of the young and the final molt which discloses the matured insect.

All insects that undergo a metamorphosis may be divided, therefore, into two classes according as the transformation from the young into the adult is direct or is completed in an intervening pupal stage. Insects of the first class are said to have *incomplete metamorphosis;* those of the second class, *complete metamorphosis.* The expressions are convenient, but misleading if taken literally, for, as we shall see, there are many degrees of "complete" metamorphosis.

The young of any insect that has a pupal stage in its life cycle is called a *larva,* and the young of an insect that does not have a pupal stage is termed a *nymph,* according to the modern custom of American entomologists. But the term "larva" was formerly applied to the immature stage of all insects, a usage which should have been preserved; and many European entomologists use the word "nymph" for the stage we call a pupa.

A larva is distinguished from a nymph by the lack of wing rudiments visible externally, and by the absence of the compound eyes. Many larvae are blind, but some of them have a group of simple eyes on each side of the head substituting for the compound eyes. Nymphs in general have the compound eyes of the adult insect, and, as seen in the young grasshopper (Fig. 9), the young dragonfly (Fig. 59), and the young cicada (Fig. 114), the nymphal wings are small pads that grow from the thoracic segments after the first or second molt. The larva, however, is not actually wingless any more than is the nymph; its wings are simply developed internally instead of externally. When the groups of cells that are destined to form the wings begin to multiply, the wing rudiments push inward instead of outward, and become small sacs invaginated into the cavity of the body, in which position they remain through all the active life of the larva. Then, at the time of the transformation, the wing sacs are everted, and appear on the outside of the pupa when the last larval skin is cast off.

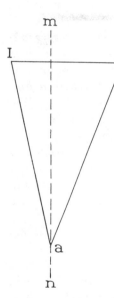

FIG. 138. Diagram of metamorphosis

If during the course of their evolution, the adult (*I*) and the larva (*L*) have independently diverged from a straight line of development (*nm*), the larva must finally attain the adult stage by a transformation (metamorphosis), the degree of which is represented by the length of the line *L* to *I*

It is difficult to discover any necessary correlation between the externally wingless condition of the larva and the existence of a pupal stage in the life of the insect; but the two for some reason go together. Perhaps it is only a coincidence. To have useless organs removed from the surface is undoubtedly an advantage to a larva, especially to such species as live in narrow spaces, or that burrow into the ground or into the stems and twigs of plants; but it probably just happened that the pupal stage was first developed in an insect that had ingrowing wings.

The typical larvae are the caterpillars, the grubs, and the maggots, young insects with little or no resemblance to their parents. The larvae of some of the beetles (Fig. 137) and of some members of the order Neuroptera, however, are much like the adults of their species, except for the lack of external wings and the compound eyes; and even among the typical larvae some species have more of the adult characters than others. The caterpillar (Fig. 136) or the grub of the May-beetle (Fig. 131 B), for example, both being provided with legs, have a much greater resemblance to an adult insect than has the wormlike legless grub of the wasp (Fig. 133 B) or the maggot of the fly (Fig. 182 D). Hence, we see, the degree of transformation may vary much even among insects that have a so-called "complete" metamorphosis.

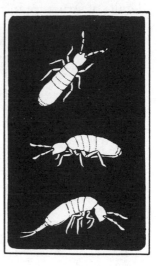

Fig. 139. Springtails, members of the Order Collembola, insects perhaps directly descended from the unknown wingless ancestors of winged insects

There are a few insects that have no metamorphosis at all. These are wingless insects belonging to the groups known as *Collembola* and *Thysanura* (Figs. 57, 139, 140) and are probably direct descendants from the primitive wingless ancestors of the winged insects. These insects during their growth shed the skin at intervals, but they do not undergo a change of form; they illustrate the normal procedure of growth by direct development from the embryo to the adult.

It must appear that the nymph, or young of an insect with incomplete metamorphosis, is merely an aberrant development of the normal form of the young as it occurs in an insect without metamorphosis. This is evident from the fact that the nymph has external wings, fully developed compound eyes, and in general the same details of structure in the legs and other parts of the body as has the adult. Most larvae, on the other hand, have few or none of the structural details of the adult that might be expected to occur in a normal postembryonic adolescent form; but they do have many characters that appear to belong to a primitive stage of evolution and that we might expect to find in an embryonic stage of development. The caterpillar, for example, has legs on the abdomen (Fig. 136, *AbL*), an embryonic feature possessed by none of the higher insects in the adult stage; it has only one claw on its thoracic legs, a character of crustaceans and myriapods, but not of adult winged insects or of nymphs. Likewise, there are certain features of the internal structure of the caterpillar that are more primitive than in any adult insect or nymph; and the same evidence of primitive or embryonic characters might be cited of other larvae.

FIG. 140. A bristletail, *Thermobia*, a member of the order Thysanura, another primitive group of wingless insects. (Twice natural size)

On the other hand, the structural details of some larvae are very much like those of the adults, and such larvae differ from the adults of their species principally in the lack of the compound eyes and of external wings.

Now, if all the insects with complete metamorphosis

[248]

have been derived from a common ancestor, as seems almost certain, then the original larvae must have been all alike, and they must have had approximately the structure of those larvae of the present time that depart least from the structure of the adult. Therefore it is evident that many larvae of the present time have somehow acquired certain embryonic characters. We may suppose, therefore, either that such larvae have had a retrogressive evolution into the embryonic stage by hatching at successively earlier ages, or that certain embryonic characters representing ancestral characters but ordinarily quickly passed over in the embryonic development, have been retained and carried on into the larval stage. The latter view seems the more probable when we consider that no larva has a purely embryonic structure, and that those larvae which have embryonic features in their anatomy present an incongruous mixture of embryonic and adult characters.

We may, therefore, finally conclude that *the larva of insects with complete metamorphosis represents the nymphal stage of insects with incomplete metamorphosis;* and that *the structure of the larva has resulted from a suppression of the peculiarly adult characters, from an invagination of the wings, a loss of the compound eyes, the retention of certain embryonic characters, and a special development of the body form and the organs suited to the particular mode of life of the larva.* By allowing for variations in all these elements that contribute to the larval make-up, except the two constants—the invagination of the wings and the loss of the compound eyes—we may account for all the variety in form and structure that the larva presents.

While, in general, the larva remains the same in structure from the time it is hatched until it transforms to the pupa, there are nearly always minor changes observable that are characteristic of its individual stages. In Chapter I we encountered the case of the little blister beetle that goes through several very different forms dur-

INSECTS

ing its development (Figs. 12, 13), and other examples of
a metamorphosis during the larval life might be given
from the other groups of insects. A larval metamorphosis
of this kind is known as *hypermetamorphosis*, and it shows
that the larva may be structurally diversified during its
growth to adapt it to several different environments or
ways of obtaining its food.

The reader was given fair warning that the subject of
insect metamorphosis would become difficult to follow,
and even now, with its realization, the writer can not
assure him that the above analysis is by any means complete or final. Much more might be said for which there
is no space here, and it is not likely that all entomologists
will accept all that has been said without a discussion,
and possibly some dissension. However, we have not
yet reached the end, for we have so far been dealing only
with the phase of metamorphosis that has produced the
nymph or the larva, and have only briefly touched upon
the reverse process which reconverts the creature into the
adult.

The pupa unquestionably has the aspect of an immature adult. It has lost all the characteristic features of
the larva, and its organs are those of the adult in the
making. It has external wing pads, legs, antennae, compound eyes. Its mouth parts are usually in a stage of
development intermediate between those of the larva and
those of the adult. Most of the pupal organs are useless,
since they are neither those of the larva nor entirely those
of the adult, and are not adapted to any special use the
pupa might make of them, except in a very few cases.
The pupa is, therefore, a helpless creature, unable to
eat, or to make any movement except by motions of the
body. It is usually said to be a "resting" stage, but its
rest is an enforced immobility, and some species attest
their impatience by an almost continuous squirming,
twisting, or wriggling of the movable parts of the body.

It is evident that it must be an advantage to the pupa

to have some kind of protection, either from the weather, or from predacious creatures that might destroy it. While most pupae are protected in one way or another, there are some that remain in exposed situations with no kind of shelter or concealment. The mosquito pupa is one of these, for it lives in the water along with the larva and floats just beneath the surface (Fig. 174 F), breathing by a pair of trumpetlike tubes that project above the surface from the anterior part of the body. The mosquito pupa is a very active creature, and can propel itself through the water, usually downward, with almost as much agility as can the larva, and by this means probably avoids its enemies. The pupa of the common lady-beetle gives another example of an unprotected pupa (Fig. 132 E). The larvae of these insects transform on the leaves where they have been feeding, and the pupae remain here attached to the leaf, unable to move except by bending the body up and down. The pupae of some of the butterflies also hang naked from the stems or leaves of plants.

The pupae of many different kinds of insects are to be found in the ground, beneath stones, under the bark of trees, or in tunnels of the leaves, twigs, or wood of plants where the larvae have spent their lives. Some of these, especially beetle pupae, are naked, soft-bodied creatures, depending on their concealment for protection. The pupae of moths and butterflies, however, are characteristically smooth, hard-shelled objects with the outlines of the legs and wings apparently sculptured on the surface (Plate 14 F). Pupae of this kind are called *chrysalides* (singular, *chrysalis*). Their dense covering is formed of a gluelike substance, exuded from the skin, that dries and forms a hard coating over the entire outer surface, binding the antennae, legs, and wings close to the body. In addition, the pupae of many moths are inclosed in a silk cocoon spun by the caterpillar. The caterpillars, as we shall learn in the next chapter, are provided with a pair of silk-producing glands which open through a hollow spine on the lower lip beneath the mouth (Fig. 155). The silk is used by the caterpillars during the feeding part of their lives in various ways, but it serves particularly for the construction of the cocoon. The most highly perfected instinct of the caterpillar is that which impels it to build the cocoon, often an intricately woven structure, just before the time of its transformation to the pupa. The caterpillar spins the cocoon around itself, then sheds its skin, which is thrust into the rear end of the cocoon as a crumpled wad. Plate 11 shows the caterpillar of a small moth that infests apple trees constructing its cocoon, finally inclosing itself within the latter, and there transforming to the pupa.

The larvae of the wasps and bees likewise inclose themselves within cocoons formed inside the cells of the comb in which they have been reared. The cocoon is made of threads, but the material is soft, and the freshly spun strands run together into a sheet that dries as a parchmentlike lining of the cell. The larvae of many of the wasplike parasitic insects that feed within the bodies of other living insects leave their hosts when ready for transformation, and spin cocoons either near the deserted host or on its body.

The maggots, or larvae, of the flies have adopted another method of acquiring protection during the pupal stage. Instead of shedding the loosened cuticula previous to the transformation, the maggot transforms within the skin, and the latter then shrinks and hardens until it becomes a tough oval capsule inclosing the larva (Fig. 182 E). The capsule is called a *puparium*. It appears, however, that the larva within the puparium undergoes another molt before it actually becomes a pupa, for, when the pupa is formed, it is found to be surrounded by a delicate membranous sheath inside the hard wall of the puparium, and when the adult fly issues it leaves this sheath and a thin pupal skin behind in the puparial shell.

[252]

PLATE 11

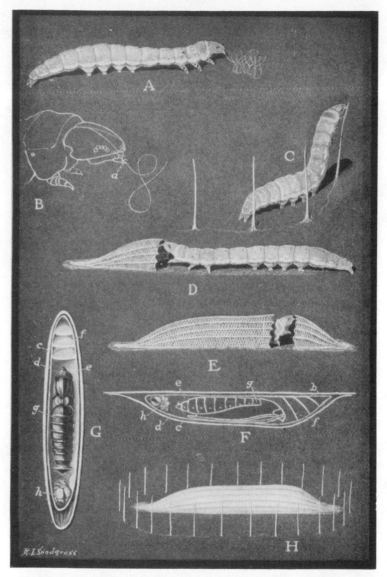

The ribbed-cocoon maker (*Bucculatrix pomifoliella*), a small caterpillar
that inhabits apple leaves

At A the caterpillar is spinning a mat of silk on the surface of a twig.
B shows the silk thread issuing from the spinneret (*a*) on the under
lip of the caterpillar. At C the caterpillar is erecting a line of silk
palisades around the site of the cocoon. D and E show the cocoon in
the course of construction, built on the silk mat. F is a diagram of
the cocoon on under surface of the support, containing the pupa (*g*)
and the shed skin of the caterpillar (*h*). G shows the interior of the
cocoon, its double walls (*c, d*), and partitions (*f*) at the front end.
H is the finished cocoon surrounded by the palisades

INSECT METAMORPHOSIS

The pupa has so many of the characters of the mature insect that we might say it is self-evident that it is a part of the adult stage, except that to say anything is "self-evident" is almost an unpardonable remark in scientific writing. However, it is clear to the eye that the pupa, in casting off the skin of the larva, has entirely discarded the larval form, except in certain insects that have a larval form in the adult stage. The pupa may retain a few unimportant larval characters, but all its principal organs are those of an adult insect in a halfway stage of development. In studying the cicada, it was observed that the adult issues from the skin of the nymph in a very immature condition. A careful dissection of a specimen at this time would show that the creature is still imperfect in many ways besides those which appear externally. By very rapid growth during the course of an hour, however, the adult form and organs are perfected. We have also noted that with insects of incomplete metamorphosis the adult is mostly formed within the nymphal skin some time before the latter is cast off. The same thing is true of a pupa. For several days before the caterpillar is ready to molt the last time, it remains almost motionless and its body contracts to perhaps less than half of the original length. The caterpillar is now said to be in a "prepupal" stage, but examination of a specimen will reveal that it has already transformed, for inside its skin is a soft pupa in a preliminary stage of development (Fig. 141 B).

This first stage of the pupa of a moth or butterfly (Fig. 141 B) is entirely comparable with the immature adult of the cicada formed inside the skin of the last stage of the nymph (Fig. 141 A). The entire pupal period, therefore, corresponds with the formative stage of the cicada, which begins within the nymphal skin and is completed about an hour after the emergence. The only external difference between the two cases is that the pupa sheds its skin, making a final added molt before it becomes a perfect insect, while the immature adult cicada goes over into the fully mature form quickly and without a molt.

We may conclude, therefore, that *the pupa of insects with complete metamorphosis corresponds with the immature stage of the adult in insects with incomplete metamorphosis.*

This idea concerning the nature of the insect pupa has been well expressed and more fully substantiated by E. Poyarkoff, and it appears to have more in its favor than the older view that the pupa corresponds with the last nymphal stage in insects with incomplete metamorphosis. According to Poyarkoff's theory, the pupa has no phylogenetic significance, that is, it does not represent any free-living stage in the evolution or ancestral history of insects; it is simply a prolonged resting period following the shedding of the last larval skin, which terminates with an added molt when the adult is fully formed.

It frequently happens that a pupa has some of the adult characters better developed than has the adult itself. The pupae of insects that

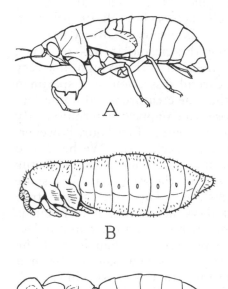

A

B

C

Fig. 141. Showing the resemblance of the pupa of an insect with complete metamorphosis to the immature adult form of an insect with incomplete metamorphosis

A, immature adult cicada, taken from the last nymphal skin. B, immature pupa of a moth, taken from the last larval skin. C, the mature pupa of a wasp

PLATE 12

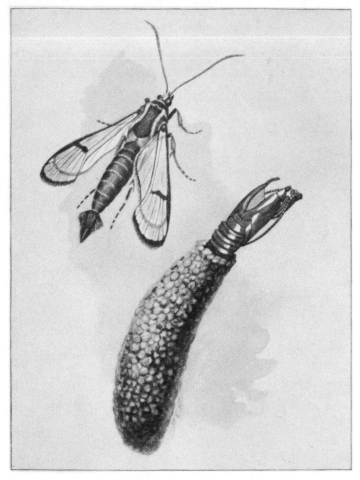

The peach-borer moth (*Aegeria exitiosa*)

Upper figure, the adult male moth (about twice natural size); lower figure, the cocoon made by the caterpillar from bits of wood, with the empty shell of the pupa projecting from the opened end

have rudimentary or shortened wings in the adult stage often have wings larger than those of the adult, indicating that the wings have been reduced in the adult since the time when the pupa was first established. Here, therefore, we see a case of metamorphosis between the pupa and the adult. Adult moths and butterflies have no mandibles or have mere rudiments of them (Fig. 163), but the jaws are often quite visible in the pupae (Fig. 159 H, *Md*), and the pupa of one moth has long, toothed mandibles which it uses to liberate itself from the cocoon before transforming to the adult.

The structural changes that accompany the transformation of the larva into an adult insect are by no means confined to the outside of the body. Much internal reorganization goes on which involves changes in the tissues themselves. The larva may have built up a highly efficient alimentary canal well adapted for handling its own particular kind of food, but perhaps the adult has adopted an entirely different diet. The alimentary canal, therefore, must be completely remodeled during the pupal stage. The nervous system and the tracheal system are often different in the larval and the adult stages, but the change in these organs is usually in the nature of a greater elaboration for the purposes of the adult, though the larva may have developed special features that are discarded.

It is in the muscles usually that the most radical reconstructive processes of the transformation from larva to adult take place. The muscles of adult insects are attached to the outer cuticular layer of the body wall, which in hard-bodied insects constitutes the "skeleton," and the mechanical differences between the larva and the adult lie in the relation between the muscles and the cuticula. With the change in the external parts between the two active stages of the insect, therefore, the larval muscles are likely to become entirely unsuited to the purposes of the adult. The special larval muscles, then, must be

[255]

cleared away, and a new muscle system must be built up suitable to the adult mechanism. Most of the other organs are transformed by a gradual replacement of cells in their tissues, with the result that each organ itself remains intact during the whole period of its alteration—the insect is never without a complete alimentary canal, its body wall always maintains a continuous surface. This condition, however, is not entirely true of the muscles, for with some insects undergoing a high degree of metamorphosis in external structure, the muscular system may suffer a complete disorganization, the fibers of the larval system being in a state of dissolution while those of the adult are in the process of development.

The muscles of adult insects, as we have just said, are attached to the outer layer of the body wall (Fig. 142). This layer is composed partly of a substance called *chitin* formed by the cellular layer of the body wall beneath it, and constitutes the cuticular skin that is shed when the insect molts. The newly-formed cuticula is soft and takes the contour of the cellular layer producing it.

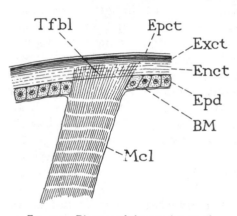

Fig. 142. Diagram of the attachment of a muscle to the body wall of an adult insect by means of the terminal fibrillae (*Tfbl*)

BM, basement membrane; *Enct*, endocuticula; *Epct*, epicuticula; *Epd*, epidermis; *Exct*, exocuticula; *Mcl*, muscle; *Tfbl*, terminal fibrillae of the muscle anchored in the cuticula

The muscles of the larva that go over into the adult stage and the new muscles of the adult must become fastened to the new cuticula, and this is possible only when the cuticula is in the soft formative stage. It has been pointed out by Poyarkoff that, for this reason, whenever new muscles are formed in an insect a new cuticula must also be produced in order that the muscle fibers may become attached to the skeleton. New muscles completed at the time of a molt may be anchored into the new cuticula formed at this time; but if the completion of the muscle tissue is delayed, the new fibers can become functional only by attaching themselves at the following molt. Conversely, if the new muscles are not perfected at the time of the last normal molt, the insect must have an extra molt later in order to give the muscles a functional connection with the body wall.

Thus Poyarkoff would explain the origin of the pupal stage in the life cycle of the insect. His theory has much to commend it, for, as Poyarkoff shows in an analysis of the various processes accompanying metamorphosis, none of the changes in any of the organs other than the muscles would seem to necessitate the production of a new cuticula and thus involve an added molt. If insects with incomplete metamorphosis add new muscles for the adult stage, such muscles must be ready-formed at the time of the last nymphal molt; but it is probable that there are few such cases in this class of insects.

Adopting Poyarkoff's theory, then, as the most plausible explanation of why a pupal stage has become separated by a molt from the fully-matured adult stage, we may say that *the reason for the pupa is probably to be found in the delayed growth of the adult muscles and in the consequent need of a new cuticula for their attachment.*

With a pupal stage once established, however, the pupa has undergone an evolution of its own, as has the larva and the adult, though to a smaller degree than either of these two active stages. The pupa is characteristically different in each of the orders of insects, and many of its features are clearly adaptations to its own mode of life.

It is one thing to know the facts and to see the meaning of metamorphosis; it is quite another to understand how it has come about that an animal undergoes a metamorphic transformation, and yet another to discover how the change is accomplished in the individual. Metamorphosis can be only a special modification of general developmental growth, and growth toward maturity by the individual goes over the same field that the species traversed in its evolution. Yet, the individual in its development may depart widely from the path of its ancestors. It may make many a detour to the right or the left; it may speed up at one place and loiter along at another; and, since the individual is rather an army of cells than a single thing, certain groups of its cells may forge ahead or go off on a bypath, while others lag behind or stop for a rest. Only one condition is mandatory, and this is that the whole army shall finally arrive at the same point at the same time. In each species, the deviations from the ancestral path, traveled for many generations, have become themselves fixed and definite trails followed by all individuals of the species. The development of the individual, therefore, may thus come to be very different from the evolutionary history of its species; and the life history of an insect with complete metamorphosis is but an extreme example of the complex course that may result when a species leaves the path of direct development to wander in the fields along the way.

The larva and the adult insect have become in many cases so divergent in structure, as a result of their separate departures from the ancestral path, that the embryo has become almost a double creature, comprising one set of cells that develop directly into the organs of the embryo and another set held in reserve to build up the adult organs at the end of the larval life. The characters of the adult are, of course, impressed upon the germ cells and must be carried over to the next generation through the embryo, but they can not be developed at the same time that the larval organs are functional. Consequently, the cells, that are to form the special tissues of the adult remain through the larval period as small groups or islands of cells in the larval tissues. These dormant cell groups are known as imaginal discs, or histoblasts. (*Imaginal* is from *imago*, an image, referring to the adult; *histoblast* means a tissue bud.)

When analyzed closely, the apparent "double" structure of the embryo will be found to be only the result of an exaggeration of the usual processes of growth, accompanied by an acceleration in certain tissues and a retardation in others. In general, wherever an adult organ is represented by an organ in the larva, even though the latter is greatly reduced, the cells that are to give this organ its adult form do not begin to develop until the larval growth is completed. But if an organ is lacking in the larval stage, the regenerative cells may start to develop at an earlier period—even in the embryo in a few cases. Hence, *the remodeling of a larval organ in the pupal stage is only a completion of that organ's normal development, and the production of a "new" organ is only the deferred development of one that has been suppressed during the larval period.*

The special organs or forms of organs that the larva has built up for its own purposes necessarily become useless when the larval life has been completed. Such organs, therefore, must be destroyed if they can not be directly made over into corresponding adult organs. Their tissues consequently undergo a process of dissolution, called *histolysis*. It can not be explained at the present time what causes histolysis, or why it begins at a certain time and in particular tissues, but histolysis is only one of the physiological processes that depend probably on the action of enzymes. In some insects a part of the degenerating tissues of the larva is devoured during the pupal stage by ameboid cells of the blood, known as *phagocytes*. It was once supposed that the phagocytes are the active agents of the destruction of the larval tissues, but this now seems improbable, since histolysis takes place whether phagocytes are present or absent.

While the larval tissues are undergoing dissolution, the adult tissues are being built up from those groups of dormant cells, the histoblasts, that have retained their vitality. Whatever it is that produces histolysis in the defunct larval tissues, it has no effect on the regenerative tissues, which now begin a period of active development, or *histogenesis* (*i.e.*, tissue building), which results in the completion of the adult organs. In most of the organs the two processes, histolysis and histogenesis, are complemental to each other, the new tissues spreading as the old are dissolved, so that there is never a lack of continuity in the parts undergoing reconstruction. It is only in the muscles, as we have already observed, that the old tissues are destroyed before the new ones are formed.

Because of the high physiological activity (*metabolism*) going on within the pupa, the blood of the insect at this stage becomes filled with a great quantity of matter resulting from the dissolution of the larval tissues. During the pupal period, the insect takes no food nor does it discharge any waste materials—the substance of the growing tissues is derived from the débris of those degenerating. But the transformation is not all direct. The insect is provided with an organ for converting some of the products of histolysis into proteid compounds that can be utilized by the tissues in histogenesis. This organ is the *fat-body* (see Chapter IV and Figure 158). During the larval life the cells of the fat body store up large quantities of fat, and in some insects glycogen, both of which energy-forming substances are discharged into the blood at the beginning of the pupal period. And now the fat cells become also active agents in the conversion of histolytic products into proteid bodies, probably by enzymes given off from their nuclei. These proteid bodies are finally also discharged into the blood, where they are absorbed as nutriment by the tissues of the newly-formed organs. At the close of the pupal period, the fat-body itself is often almost entirely consumed or

PLATE 13

The red-humped caterpillar (*Schizura concinna*)

A, the moth in position of repose (natural size). B, moth with wings spread. C, under surface of apple leaf, showing eggs at *a*, and young caterpillars feeding at *b*. D, a caterpillar in next to last stage of growth. E, full-grown caterpillars (one-half larger than natural size). F, two cocoons on ground among grass and dead leaves, one cut open showing caterpillar within before transforming to pupa

is reduced to a few scattered cells, which build up the fat-body of the adult.

The internal adult organs undergo a continuous development throughout the pupal period and are practically complete when the latter terminates with the molt to the adult. But the external parts, after quickly attaining a halfway stage of development at the beginning of the pupal metamorphosis are checked in their growth by the hardening of the cuticular covering of the body wall, and in their half-formed shape they must remain to the end of the pupal period. It is only by a subsequent growth of the cellular layer of the body wall beneath the loosened cuticula of the pupa that the external adult parts are finally perfected in structure; and it is only when the pupal cuticula is then cast off and the organs cramped within it are given freedom to expand that the adult insect at last appears in its fully mature form.

CHAPTER IX

THE CATERPILLAR AND THE MOTH

The Life of a Caterpillar

IT is one of those bleak days of early spring that so often follow a period of warmth and sunshine, when living things seem led to believe the fine weather has come to stay.

Out in the woods a band of little caterpillars is clinging to the surface of something that appears to be an oval swelling near the end of a twig on a wild cherry tree (Fig. 143). The tiny creatures, scarce a tenth of an inch in length, sit motionless, benumbed by the cold, many with bodies bent into half circles as if too nearly frozen to straighten out. Probably, however, they are all unconscious and suffering nothing. Yet, if they were capable of it, they would be wondering what fate brought them into such a forbidding world.

But fate in this case was disguised most likely in the warmth of yesterday, which induced the caterpillars to leave the eggs in which they had safely passed the winter. The empty eggshells are inside the spindle-shaped thing that looks so like a swelling of the twig, for in fact this is merely a protective covering over a mass of eggs glued fast to the bark. The surface of the covering is perforated by many little holes from which the caterpillars emerged, and is swathed in a network of fine silk threads which the caterpillars spun over it to give themselves a surer footing and one they might cling to unconsciously in the event of adverse weather, such as that which makes them helpless now. When nature designs any creature

PLATE 14

The tent caterpillar (*Malacosoma americana*)

A, an egg mass on an apple twig (about natural size). B, young caterpillar feeding on an opening leaf bud. C, branches of an apple tree with a tent in a fork, from which trails of silk lead outward to the twigs where the caterpillars are feeding on the leaves. D, a full-grown caterpillar (three-fourths natural size). E, cocoon. F, pupa, taken from a cocoon. G, male moth. H, female moth laying eggs

to live under trying circumstances she grants it some safeguard against destruction.

The web-spinning habit is one which, as we shall see, these caterpillars will develop to a much greater extent later in their lives, for our little acquaintances are young tent caterpillars. They are found most often among woodland trees, on the chokecherry and the wild black cherry. But they commonly infest apple trees in the orchards, and for this reason their species has been named the apple-tree tent caterpillar, to distinguish it from related forms that do not commonly inhabit cultivated fruit trees. The scientific name is *Malacosoma americana.*

The egg masses of the tent caterpillar moths are not hard to find at this season. They are generally placed near the tips of the twigs, which they appear to surround, and being of the same brownish color as the bark, they look like swollen parts of the twigs themselves (Plate 14 A, Fig. 144 A). Most of them are five-eighths to seven-eighths of an inch in length and almost half of this in width, but they vary in thickness with the diameter of

Fig. 143. Young tent caterpillars on the egg mass from which they have just hatched. (1¼ times natural size)

the twig. A closer inspection shows that the mass really clasps the twig, or incloses it like a thick jacket lapped clear around. In form the masses are usually symmetrical, tapering at each end, but some are of irregular shapes, and those that have been placed at a forking or against a bud have one end enlarged.

The greater part of an egg mass consists of the covering material, which is a brittle, filmy substance like dry mucilage. Some of it is often broken away, and sometimes the tops of the eggs are entirely bare. The eggs are placed in a single layer next the bark (Fig. 144 B), and there are usually 300 or 400 of them. They look like little, pale-gray porcelain jars packed closely together and glued to the twig by their rounded and somewhat compressed lower ends. The tops are flat or a little convex. Each egg is the twenty-fourth of an inch in height, about two-thirds of this in width, and has a capacity of one caterpillar. The covering is usually half again as deep as the height of the eggs, but varies in thickness in different specimens. The outer surface is smooth and polished, but the interior is full of irregular, many-sided air spaces, separated from one another by thin partitions (B).

Wherever the covering of an egg mass has been broken away, the bases of the partition walls leave brown lines that look like cords strapped and tied into an irregular net over the eggs (B), as if for double security against insurrection on the part of the inmates. But neither shells nor fastenings will offer effective resistance to the little caterpillars when they are taken with the urge for freedom. Each is provided with efficient cutting instruments in the form of sharp-toothed jaws that will enable it to open a round hole through the roof of its cell (Fig. 144 C). The superstructure is then easily penetrated, and the emerging caterpillar finds itself on the surface of its former prison, along with several hundred brothers and sisters when all are out.

All this time the members of that unfortunate brood we noted first have been clinging benumbed and motionless to the silk network on the covering of their deserted eggs. The cold continues, the clouds are threatening, and during the afternoon the hapless creatures are drenched by hard and chilling rains. Through the night following they are tossed in a northwest gale, while the temperature drops below freezing. The next day the wind continues, and frost comes again at night. For three days the caterpillars endure the hostility of the elements, without food, without shelter. But already the buds on the cherry tree are sending out long green points, and when the temperature moderates on the fourth day and

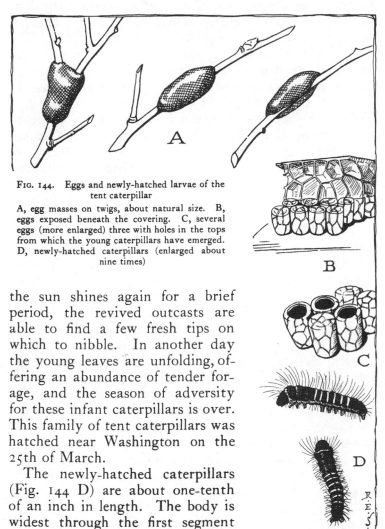

FIG. 144. Eggs and newly-hatched larvae of the tent caterpillar

A, egg masses on twigs, about natural size. B, eggs exposed beneath the covering. C, several eggs (more enlarged) three with holes in the tops from which the young caterpillars have emerged. D, newly-hatched caterpillars (enlarged about nine times)

the sun shines again for a brief period, the revived outcasts are able to find a few fresh tips on which to nibble. In another day the young leaves are unfolding, offering an abundance of tender forage, and the season of adversity for these infant caterpillars is over. This family of tent caterpillars was hatched near Washington on the 25th of March.

The newly-hatched caterpillars (Fig. 144 D) are about one-tenth of an inch in length. The body is widest through the first segment

[265]

and tapers somewhat toward the other end. The general color is blackish, but there is a pale gray collar on the first segment back of the head and a grayish line along the sides of the body. Most of the segments have pale rear margins above, which are often bright yellow or orange on the fourth to the seventh segments. There is usually a darker line along the middle of the back. The body is covered with long gray hairs, those on the sides spreading outward, those on the back curving forward. After a few days of feeding the caterpillars increase to nearly twice their length at hatching.

When the weather continues fair after the time of hatching, the caterpillars begin their lives with happier days, and their early history is different from that of those unfortunates described above. Three other broods, which were found hatching on March 22, before the period of bad weather had begun, were brought indoors and reared under more favorable circumstances. These caterpillars spent but little time on the egg masses and wasted only a few strands of silk upon them. They were soon off on exploring expeditions, small processions going outward on the twigs leading from the eggs or their vicinity, while some individuals dropped at the ends of threads to see what might be below. Most, however, at first went upward, as if they knew the opening leaf buds should lie in that direction. If this course, though, happened to lead them up a barren spur, a squirming, furry mob would collect on the summit, apparently bewildered by the trick their instinct had played upon them. On the other hand, many followed those that first dropped down on threads, these in turn adding other strands till soon a silken stairway was constructed on which individuals or masses of little woolly bodies dangled and twisted, as if either enjoying the sport or too fearful to go farther.

For several days the young caterpillars led this happy, irresponsible life, exploring twigs, feeding wherever an open leaf bud was encountered, dangling in loose webs, but spinning threads everywhere. Yet, in each brood, the individuals kept within reach of one another, and the trails of silk leading back to the main branch always insured the possibility of a family reunion whenever this should be desired.

One morning, the 27th, one family had gathered in its scattered members and these had already spun a little tentlike web in the crotch between the main stem of the supporting twig and two small branches (Fig. 145). Some members were crawling on the surface of the tent, others were resting within, still others were traveling back and forth on the silk trails leading outward on the branches, and the rest were massed about the buds devouring the young leaves. The establishment of the tent marks the beginning of a change in the caterpillars' lives; it entails responsibilities that demand a fixed course of daily living. In the lives of the tent caterpillars this point is what the beginning of school days is to us—the end of irresponsible freedom, and the beginning of subjection to conventional routine.

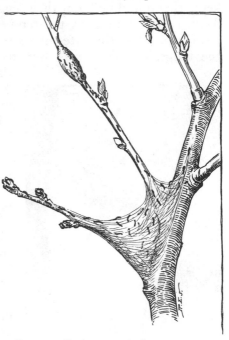

Fig. 145.   First tent made by young tent caterpillars.   (About half natural size)

Every tent caterpillar family that survives infancy eventually reaches the point where it begins the construction of a tent, but the early days are not always spent alike, even under similar circumstances, nor is the tent always begun in the same manner.

In the State of Connecticut, where the season for both plants and insects is much later than in the latitude of Washington, three broods of tent caterpillars were observed hatching on April 8 of the same year. These caterpillars also met with dull and chilly weather that kept them huddled on their egg coverings for several days. After four days the temperature moderated sufficiently to allow the caterpillars to move about a little on the twigs, but none was seen feeding till the 14th—six days after the hatching. Yet they had increased in size to about one-eighth of an inch in length.

Wherever these caterpillars camped in their wanderings over the small apple trees they inhabited, they spun a carpet of silk to rest upon, and there the whole family collected in such a crowded mass that it looked like a round, furry mat (Fig. 146). The carpets afforded the caterpillars a much safer bed than the bare, wet bark of the tree, for if the sleepers should become stupefied by cold the claws of their feet would mechanically hold them fast to the silk during the period of their helplessness. The test came on the 16th and the night following, when the campers were soaked by hard, cold rains

Fig. 146. Young tent caterpillars matted on a flat sheet of web spun in the crotch between two branches. (About natural size)

[268]

till they became so inert they seemed reduced to lifeless masses of soggy wool. On the afternoon of the 17th the temperature moderated, the sun came out a few times, the wetness evaporated from the trees, and most of the caterpillars revived sufficiently to move about a little and dry their fur. Though a few had been washed off the carpets by the violence of the storm and had perished on the ground, and in one camp about twenty dead were left behind on the web, the majority had survived.

For several days after this, during better weather, the caterpillars of these families continued their free existence, feeding at large on the opening buds, but returning during resting periods to the webs, or constructing new ones at more convenient places. Often each family split into several bands, each with its own retreat, yet all remained in communication by means of the silk trails the members left wherever they went.

The camping sites were either against the surface of a branch or in the hollow of a crotch. Though the carpet-like webs stretched over these places were spun apparently only to give secure footing, those at the crotches often roofed over a space well protected beneath, and frequently many of the caterpillars crawled into these spaces to avail themselves of their shelter. Yet for twelve days none of the broods constructed webs designed for coverings. Then, on the morning of the 20th, one family was found to have spun several sheets of silk above the carpet on which its members had rested for a week, and all were now inside their first tent. These caterpillars were nearing the end of their first stage, and two days later the first molted skins were found in the tent, fourteen days after the date of hatching.

In Stage II the caterpillars have a new color pattern and one which begins to suggest that characteristic of the species in its more mature stages (Fig. 148). On the upper part of the sides the dark color is broken into a series of quadrate spots each spot partially split lengthwise by a light streak, and the whole series on each side is bordered above and below by distinct pale lines, the upper line often yellowish. Below the lower line there is a dark band, and below this another pale line just above the bases of the legs. The back of the first body segment has a brown transverse shield, and the last three segments are continuously brown, without spots or lines.

From now on the tents increase rapidly in size by successive additions of web spun over the tops and sides, each new sheet covering a flat space between itself and the last. The old roofs thus become successively the floors of the new stories. The latter, of course, lap over on the sides, and many continue clear around and beneath the original structure; but since the tent was started in a crotch, the principal growth is upward with a continual expansion at the top. During the building period a symmetrical tent is really a beautiful object (Plate 14 C). Half hidden among the leaves, its silvery whiteness pleasingly contrasts with the green of the foliage; its smooth silk walls glisten where the sun falls upon them and reflect warm grays and purples from their shadows.

The caterpillars have adopted now a community form of living; all feed together, all rest and digest at the same time, all work at the same time, and their days are divided into definite periods for each of their several duties. There is, however, no visible system of government or regulation, but with caterpillars acts are probably functions; that is, the urge probably comes from some physiological process going on within them, which may be influenced somewhat by the weather.

The activities of the day begin with breakfast. Early in the morning the family assembles on the tent roof, and about six-thirty, proceeds outward in one or several orderly columns on the branches. The leaves on the terminal twigs furnish the material for the meal. After two hours or more of feeding, appetites are appeased, and the caterpillars go back to the surface of the tent, usually by eight-thirty or nine o'clock. Here they do a little spinning on its walls, but no strenuous work is attempted at

Fig. 147. Mature tent caterpillars feeding at night

this time, and generally within half an hour the entire family is reassembled inside the tent. Most frequently the crowd collects first in the shady side of the outermost story, but as the morning advances the caterpillars seek the cooler inner chambers, where they remain hidden from view.

In the early part of the afternoon a light lunch is taken. The usual hour is one o'clock, but there is no set time. Occasionally the participants appear shortly after eleven, sometimes at noon, and again not until two or three o'clock, and rarely as late as four. As they assemble on the roof of the tent they spin and weave again until all are ready to proceed to the feeding grounds. This meal lasts about an hour. When the caterpillars return to the tent they do a little more spinning before they retire for the afternoon siesta. Luncheon is not always fully attended and is more popular with caterpillars in the younger stages, being dispensed with entirely, as we shall see, in the last stage.

Dinner, in the evening, is the principal meal of the day, and again there is much variation in the time of service. Daily observations made on five Connecticut colonies from the 8th to the 26th of May gave six-thirty p.m. as the earliest record for the start of the evening feeding, and nine o'clock as the latest; but the dinner hour is preceded by a great activity of the prospective diners assembled on the outsides of the tents. Though the energy of the tent caterpillars is never excessive, it appears to reach its highest expression at this time. The tent roofs are covered with restless throngs, most of the individuals busily occupied with the weaving of new web, working apparently in desperate haste as if a certain task had been set for them to finish before they should be allowed to eat. Possibly, though, the stimulus comes merely from a congestion of the silk reservoirs in their bodies, and the spinning of the thread affords relief.

The tent caterpillar does not weave its web in regular loops of thread laid on by a methodical swinging of the head from side to side, which is the method of most caterpillars. It bends the entire body to one side, attaches the thread as far back as it can reach, then runs forward a few paces and repeats the movement, sometimes on the same side, sometimes on the other. The direction in which the thread is carried, however, is a haphazard one, depending on the obstruction the spinner meets from others working in the same manner. Among the crowd of weavers there are always a few individuals that are not working, though they are just as active as the others. These are running back and forth over the surface of the tent, like boarders impatiently awaiting the sound of the dinner bell. Perhaps they are in-dividuals that have finished their work by exhausting their supply of silk.

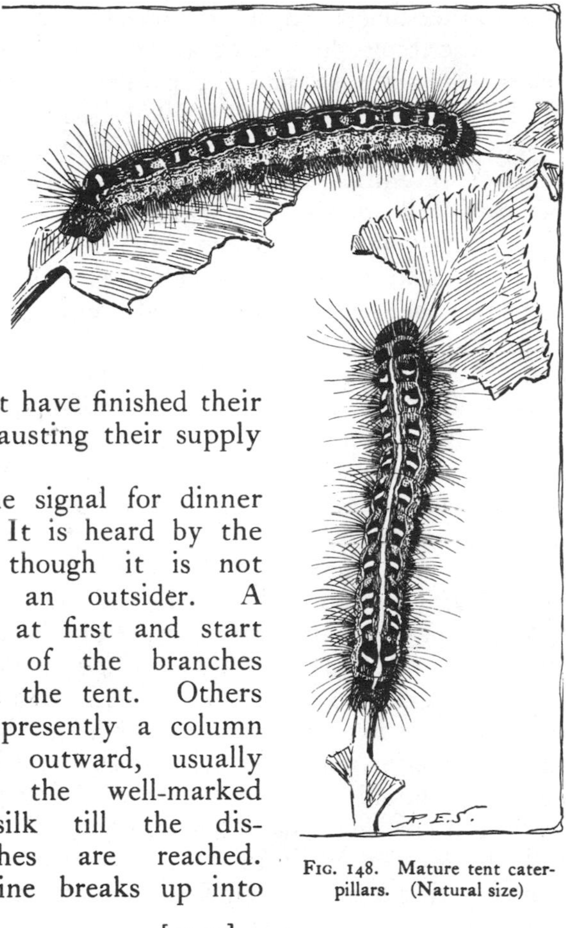

At last the signal for dinner is sounded. It is heard by the caterpillars, though it is not audible to an outsider. A few respond at first and start off on one of the branches leading from the tent. Others follow, and presently a column is marching outward, usually keeping to the well-marked paths of silk till the dis-tant branches are reached. Here the line breaks up into

FIG. 148. Mature tent cater-pillars. (Natural size)

[273]

several sections which spread out over the foliage. The tent is soon deserted. For one, two, or three hours the repast continues, the diners often returning home late at night. Observations indicate that this is the regular habit of the tent caterpillar in its earlier stages, and perhaps up to the sixth or last stage of its life. In at least nine instances the writer noted entire colonies back in the tent for the night at hours ranging from nine to eleven p.m.; but sometimes a part of the crowd was still feeding when last observed.

In describing the life of a community of insects it is seldom possible to make general statements that will apply to all the individuals. The best that a writer can do is to say what he sees most of the insects do, for, as in other communities, there are always those eccentric members who will not conform with the customs of the majority. Occasionally a solitary tent caterpillar may be seen feeding between regular mealtimes. Often one works alone on the tent, spinning and weaving long after its companions have quit and gone below for the midday rest. Such a one appears to be afflicted with an over-developed sense of responsibility. Then, too, there is nearly always one among the group in the tent who can not get to sleep. He flops this way and that, striking his companions on either side and keeping them awake also. These are annoyed, but they do not retaliate; they seem to realize that their restless comrade has but a common caterpillar affliction and must be endured.

Many of these little traits make the caterpillars seem almost human. But, of course, this is just a popular form of expression; in fact, it expresses an idea too popular —we take an over amount of satisfaction in referring to our faults as particularly human characteristics. What we really should say is not how much tent caterpillars are like us in their shortcomings, but how much we are still like tent caterpillars. We both revert more or less in our instincts to times before we lived in communities, to times when our ancestors lived as individuals irresponsible one to another.

The tent caterpillars ordinarily shed their skins six times during their lives. At each molt the skin splits along the middle of the back on the first three body segments and around the back of the head. It is then pushed off over the rear end of the body, usually in one piece, though most other caterpillars cast off the head covering separate from the skin of the body in all molts but the last. The moltings take place in the tent, except the molt of the caterpillar to the pupa, and each molt renders the caterpillars inactive for the greater part of two days. When most of them shed their skins at the same time there results an abrupt cessation of activity in the colony. By the time the caterpillars reach maturity the discarded skins in a tent outnumber the caterpillars five to one.

The first stage of the caterpillars, as already described (Fig. 144 D), suggests nothing of the color pattern of the later stages, but in Stage II the spots and stripes of the mature caterpillars begin to be formed. In succeeding stages the characters become more and more like those of the sixth or last stage (Plate 14 D, Fig. 148), when the colors are most intensified and their pattern best defined. Particularly striking now are the velvety black head with the gray collar behind; the black shield of the first segment split with a medium zone of brown; the white stripe down the middle of the back; the large black lateral blotches, each inclosing a spot of silvery bluish white; the distinctly bluish color between and below the blotches; and the hump on the eleventh segment, where the median white line is almost obliterated by the crowding of the black from the sides. Yet the creatures wearing all this lavishness of decoration make no ostentatious show, for the colors are all nicely subdued beneath the long reddish-brown hairs that clothe the body. In the last stage, the average full-grown caterpillar is about two inches long, but some reach a length of two and a half inches when fully stretched out.

In Connecticut, the tent caterpillars begin to go into their sixth and last stage about the middle of May. They now change their habits in many ways, disregarding the conventionalities and refusing the responsibilities that bound them in their earlier stages. They do little if any spinning on the tent, not even keeping it in decent repair. They stay out all night to feed (Fig. 147), unless adverse weather interferes, thus merging dinner into breakfast in one long nocturnal repast. This is attested by observations made through most of several nights, when the caterpillars of four colonies which went out at the usual time in the evenings were found feeding till at least four o'clock the following mornings, but were always back in the tents at seven-thirty a.m. When the caterpillars begin these all-night banquets, they dispense with the midday lunch, their crops being so crammed with food by morning that the entire day is required for its digestion. Some writers have described the tent caterpillars as nocturnal feeders, and some have said they feed three times a day. Both statements, it appears, are correct, but the observers have not noted that the two habits pertain to different periods of the caterpillars' history.

At any time during the caterpillars' lives adverse weather conditions may upset their daily routine. For two weeks during May, days and nights had been fair and generally warm, but on the 17th the temperature did not get above 65° F., and in the afternoon threatening clouds covered the sky. In the evening light rains fell, but the caterpillars of the five colonies under observation came out as usual for dinner and were still feeding when last observed at nine p.m. Rains continued through the night, however, and the temperature stood almost stationary between 50° and 55°.

The next morning three of the small trees containing the colonies were festooned with water-soaked caterpillars, all hanging motionless from leaves, petioles, and twigs, be-numbed with exposure and incapable of action—more miserable-looking insects could not be imagined. No in-stinct of protection, apparently, had prevailed over their appetites; till at last, overcome by wet and cold, they were saved only by some impulse that led them to grasp the support so firmly with the abdominal feet that they hung there mechanically when senses and power of move-ment were gone. Some clung by the hindmost pair of feet only, others grasped the support with all the abdominal feet. One colony and most of another were safely housed in their tents. These had evidently retreated before helplessness overtook them.

By eight o'clock in the morning many of the suspended caterpillars were sufficiently revived to resume activity. Some fed a little, others crawled feebly toward the tents. By 9:45 most were on their way home, and at 10:45 all were under shelter.

Gentle rains fell during most of the day, but the tem-perature gradually rose to a maximum of 65°. Only a few caterpillars from the youngest colony came out to feed at noon. In the evening there was a hard, drenching rain, after which several caterpillars from two of the tents ap-peared for dinner. The next morning, the 19th, the tem-perature dropped to 49°, light rains continued, and not a caterpillar from any colony ventured out for breakfast. It looked as if they had learned their lesson; but it is more probable they were simply too cold and stiff to leave the tents. In the afternoon the sky cleared, the temperature rose, and the colonies resumed their normal life.

The tent caterpillars' mode of feeding is to devour the leaves clear down to the midribs (Figs. 148, 149), and in this fashion they denude whole branches of the trees they inhabit. Since the caterpillars have big appetites, it some-times happens that a large colony in a small tree or several colonies in the same tree may strip the tree bare before they reach maturity. The writer never saw a colony reduced to this extremity by its own feeding, but produced similar conditions for one in a small apple tree by removing all the leaves. This was on May 19, and the caterpillars were mostly in their fifth stage. At seven o'clock in the evening the caterpillars in this colony came out as usual, and, after doing the customary spinning on the tent, started off to get their dinner, suspecting nothing till they came to the cut-off ends of the branches. Then they were clearly bewildered —they returned and tried the course over again; they tried another branch, all the other branches; but all ended alike in bare stumps. Yet there were the accustomed trails, and their instincts clearly said that silk paths led to food. So all night the caterpillars hunted for the missing leaves; they went over and over the same courses, but none ventured below the upper part of the trunk. By 3:45 in the morning

FIG. 149. Twigs of choke cherry and of apple denuded by tent caterpillars

many had given up and had gone back to the tent, but the rest continued the hopeless search. At seven-thirty a few bold explorers had discovered some remnants of water sprouts at the base of the tree and fed there till ten o'clock. At eleven all were back in the tent.

At two o'clock in the afternoon the crowd was out again and a mass meeting was being held at the base of the tree. But nobody seemed to have any idea of what to do, and no leader rose to the occasion. A few cautious scouts were making investigations over the ground to the extent of a foot or a little more from the base of the trunk, but, though there were small apple trees on three sides five feet away, only one small caterpillar ventured off toward one of these. He, however, missed the mark by twelve inches and continued onward; but probably chance eventually rewarded him. At three p.m. the meeting broke up, and the members went home. They were not seen again that evening or the next morning.

During this day, the 21st, and the next, an occasional caterpillar came out of the tent but soon returned, and it was not till the evening of the 22nd that a large number appeared. These once more explored the naked branches and traveled up and down the new paths on the trunk, but none was observed to leave the tree. On the 23rd and 24th no caterpillars were seen. On the 25th the tent was opened and only two small individuals were found within it. Each of these was weak and flabby, its alimentary canal completely empty. But what had become of the rest? Probably they had wandered off unobserved one by one. Certainly there had been no organized migration. Solitary caterpillars were subsequently found on a dozen or more small apple trees in the immediate vicinity. It is likely that most of these had molted and had gone into the last stage, since their time was ripe, but this was not determined.

After the caterpillars go over into their last stage, the tents are neglected and rapidly fall into a state of dilapidation. Birds often poke holes in them with their bills and rip off sheets of silk which they carry away for nest-building purposes. The caterpillars do not even repair these damages. The rooms of the tent become filled with accumulations of frass, molted skins, and the shriveled bodies of dead caterpillars. The walls are discolored by rains which beat into the openings and soak through the refuse. Thus, what were shapely objects of glistening silk are transmuted into formless masses of dirty rags.

But the caterpillars, now in their finest dress, are oblivious of their sordid surroundings and sleep all day amidst these disgusting and apparently insanitary conditions. However, the life in the tents will soon be over; so it appears the caterpillars simply think, "What's the use?" But of course caterpillars do not think; they arrive at results by instinct, in this case by the lack of an instinct, for they have no impulse to keep the tents clean or in repair when doing so would be energy wasted. Nature demands a practical reason for most things.

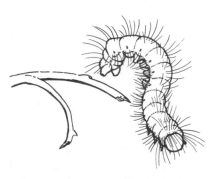

FIG. 150. A tent caterpillar in the last stage of its growth, leaving the tree containing its nest by jumping from the end of a twig to the ground

The tent life continues about a week after the last molt, and then the family begins to break up, the members leaving singly or in bands, but always as individuals without further concern for one another. Judging from their previous methodical habits, one would suppose that the caterpillars starting off on their journeys would simply go down the trunks of the trees and walk away. But no; once in their life they must have a dramatic moment. A caterpillar comes rushing out of a tent as if suddenly awakened from some terrible dream or as if pursued by a demon, hurries outward along a branch, goes to the end of a spur or the tip of a leaf, and without slackening continues into space till the end of the support tickles his stomach, when suddenly he gives a flip into the air, turns a somersault, and lands on the ground (Fig. 150).

The first performance of this sort was observed on May 15 in the Connecticut colonies. On the afternoon of the 19th, twenty or more caterpillars from two neighboring colonies were seen leaving the trees in the same fashion within half an hour. Most of the members of one of these colonies had their last molt on May 12 and 13. During the next few days other caterpillars were observed jumping from four trees containing colonies under observation. All of these went off individually at various times, but most of them early in the afternoon. Many caterpillars simply drop off when they reach the end of the branch, without the acrobatic touch, but only three were seen to go down the trunk of a tree in commonplace style.

The population of the tents gradually decreases during several days following the time when the first caterpillar departs. One of the two tents from which the general exodus was noted on May 19 was opened on the 21st and was found to contain only one remaining caterpillar. On the evening of the 22nd a solitary individual was out feeding from the other tent. The two younger colonies maintained their numbers until the 22nd, after which they diminished till, within a few days, their tents also were deserted. The members of all these colonies hatched from the eggs on April 8, 9, and 10, so seven weeks is the greatest length of time that any of them spent on the trees of their birth. The caterpillar that left the tent on the 15th came from a colony that began to hatch on April 10, giving an observed minimum of thirty-six days.

After the mature caterpillars leave the tents, they wander at large and feed wherever they find suitable provender, enjoying for a while a new life free from the domestic routine that has bound them since the days of their infancy. But even their liberty has an ulterior purpose: the time is now approaching when their lives as caterpillars must end and the creatures must go through the mysteries of transformation, which, if successfully accomplished, will convert them into winged moths. It would clearly be most unwise for the caterpillars of a colony to undergo the period of their metamorphosis huddled in the remains of the tent, where some untoward event might bring destruction to them all. Nature has, therefore, implanted in the tent caterpillar a migratory urge, which now becomes active and leads the members of

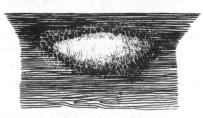

a family to scatter far and wide. About a week is allowed for the dispersal, and then, as each wanderer feels within the first warnings of approaching dissolution, it selects a suitable place for inclosing itself in a cocoon.

FIG. 151. The cocoon of a tent caterpillar. (Natural size)

It is difficult to find many cocoons in the neighborhood where large numbers of caterpillars have dispersed, but such as may be recovered will be found among blades of grass, under ledges of fences, or in sheds and barns where they are not disturbed. The cocoon is a slender oval or almost spindle-shaped object, the larger ones being about an inch long and half an inch in width at the middle (Plate 14 E, Fig. 151). The structure is spun of white silk thread, but its walls are stiffened and colored by a yellowish substance infiltrated like starch through the meshes of the fabric.

In building the cocoon the caterpillar first spins a loose network of threads at the place selected, and then, using this for a support, weaves about itself the walls of the final structure. On account of its large size, as compared with the size of the cocoon, the caterpillar is forced to double on itself to fit its self-imposed cell. Most of its hairs, however, are brushed off and become interlaced with the threads to form a part of the cocoon fabric. When the spinning is finished, the caterpillar ejects a yellowish, pasty liquid from its intestine, which it smears all over the inner surface of the case; but the substance spreads through the meshes of the silk, where it quickly dries and gives the starchy stiffness to the walls of the finished cocoon. It readily crumbles into a yellow powder, which becomes dusted over the caterpillar within and floats off in a small yellow cloud whenever a cocoon is pulled loose from its attachments.

The cocoon is the last resting place of the caterpillar. If the insect lives, it will come out of its prison as a moth, leaving the garments of the worm behind. It may, however, be attacked by parasites that will shortly bring about its destruction. But even if it goes through the period of change successfully it must remain in the cocoon about three weeks. In the meantime it will be of interest to learn something of the structure of a caterpillar, the better to understand some of the details of the process of its transformation.

The Structure and Physiology of the Caterpillar

A caterpillar is a young moth that has carried the idea of the independence of youth to an extreme degree, but which, instead of rising superior to its parents, has degenerated into the form of a worm. An excellent theme this would furnish to those who at present are bewailing what they believe to be a shocking tendency toward an excess of independence on the part of the young of the human species; but the moral aspect of the lesson somewhat loses its force when we learn that this freedom of the caterpillar from parental restraint gives advantages to both young and adults and therefore results in good to the species as a whole.   Independence entails responsibilities.   A creature that leaves the beaten paths of its ancestors must learn to take care of itself in a new way.   And

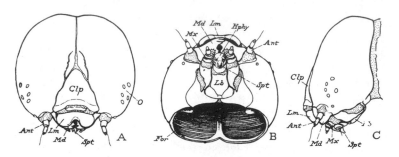

FIG. 152.   The head of a tent caterpillar

A, facial view. B, under surface. C, side view. *Ant*, antenna; *Clp*, clypeus; *For*, opening of back of head into body; *Hphy*, hypopharynx; *Lb*, labium; *Lm*, labrum; *Md*, mandible; *Mx*, maxilla; *O*, eyes; *Spt*, spinneret

this the caterpillar has learned to do preeminently well, as it has come up the long road of evolution, till now it possesses both instincts and physical organs that make it one of the dominant forms of insect life.

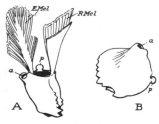

FIG. 153.   The mandibles, or biting jaws, of the tent caterpillar detached from the head

A, front view of right mandible. B, under side of the left mandible. *a* and *p*, the anterior socket and posterior knob by which the jaw is hinged to the head; *EMcl*, *RMcl*, abductor and adductor muscles that move the jaw in a transverse plane

The external organs of principal interest in the caterpillar are those of the head (Fig. 152). These include the eyes, the antennae, the mouth, the jaws, and the silk-spinning instrument. A facial view of a caterpillar's head shows two large, hemispherical lateral areas separated by a median suture above and a triangular plate (*Clp*) below. The walls of the lateral hemispheres give attachment to the muscles that move the jaws, and their size is no index of the brain-power of

[284]

the caterpillar, since the insect's brain occupies but a small part of the interior of the head (Fig. 154, *Br*). From the lower edge of the triangular facial plate (Fig. 152, *Clp*) is suspended the broad, notched front lip, or labrum (*Lm*) that hangs as a protective flap over the bases of the jaws. At the sides of the labrum are the very small antennae (*Ant*) of the caterpillar. On the lower part of each lateral hemisphere of the head are six small simple eyes, or *ocelli* (*O*), five in an upper group, and one near the base of the antenna. With all its eyes, however, the caterpillar

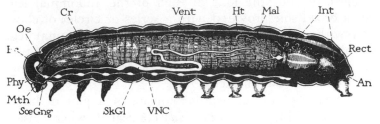

FIG. 154. Diagrammatic lengthwise section of a caterpillar, showing the principal internal organs, except the tracheal system

An, anus; *Br*, brain; *Cr*, crop; *Ht*, heart; *Int*, intestine; *Mal*, Malpighian tubule (two others are cut off near their bases); *Mth*, mouth; *Oe*, oesophagus; *Phy*, pharynx; *Rect*, rectum; *SkGl*, silk gland; *SoeGng*, suboesophageal ganglion; *Vent*, stomach, or ventriculus; *VNC*, ventral nerve cord

appears to be very nearsighted and gives little evidence of being able to distinguish more than the presence or absence of an object before it, or the difference between light and darkness. Those tent caterpillars that were starving on the denuded tree failed to perceive other food trees in full leaf only a few feet away.

The general external form and structure of the tent caterpillar is shown at A of Figure 159. The body is soft and cylindrical. The head is a small, hard-walled capsule attached to the body by a short flexible neck. Back of the head and neck comes first a body region consisting of three segments that bear each a pair of small, jointed legs (*L*); and then comes a long region composed of ten segments supported on five pairs of short, unjointed legs

(*AL*), the first four pairs being on the third, fourth, fifth and sixth segments, and the last on the tenth segment. The region of the three segments in the caterpillar bearing the jointed legs corresponds with the thorax of an adult insect (Fig. 63, *Th*), and that following corresponds with the abdomen (*Ab*). The thorax of the adult insect constitutes the locomotor center of the body, but the wormlike caterpillar has no special locomotor region, and hence its body is not separated into thorax and abdomen. The thoracic legs of the caterpillar terminate each in a single claw, but the foot of each of the abdominal legs has a broad sole provided with a series or circlet of claws and with a central vacuum cup. The abdominal legs of the caterpillar, therefore, are important organs of progression, and are the chief organs of grasping or of clinging to hard or flat surfaces.

The jaws of the caterpillar consist of a pair of large, strong mandibles (Fig. 152, *Md*) concealed, when closed, behind the labrum. Each jaw is hinged to the lower edge of the cranium at the side of the mouth by two ball-and-socket hinges in such a manner that, when in action, it swings outward and inward on a lengthwise axis. The cutting edges are provided with a number of strong teeth (Fig. 153), the points of which come together or slide past each other when the jaws are closed.

The large complex organ that projects behind or below the mouth like a thick under lip (Fig. 152 C) is a combination of three parts that are separate in other insects. These are the second pair of soft jaw appendages, called *maxillae* (B, C, *Mx*), and the true under lip, or *labium* (*Lb*). The most important part of this composite structure in the caterpillar, however, is a hollow spine (A, B, C, *Spt*) pointed downward and backward from the end of the labium. This is the *spinneret*. From it issues the silk thread with which the caterpillar weaves its tent and its cocoon.

The fresh silk is a liquid formed in two long, tubular glands extending far back from the head into the body of the caterpillar (Fig. 154, *SkGl*). The middle part of each tube is enlarged to serve as a reservoir where the silk liquid may accumulate (Fig. 155 A, *Res*); the anterior narrowed part constitutes the duct (*Dct*), and the ducts

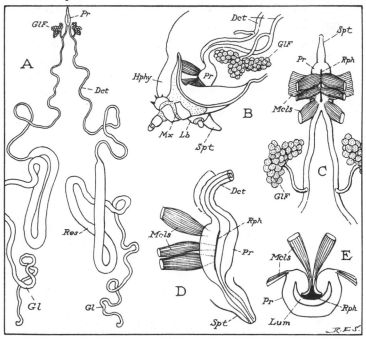

FIG. 155.   The silk glands and spinning organs of the tent caterpillar

A, the silk-forming organs, consisting of a pair of tubular glands (*Gl, Gl*), each enlarging into a reservoir (*Res*), and opening through a long duct (*Dct*) into the silk press (*Pr*), with a pair of accessory glands (glands of Filippi, *GlF*) opening into the ducts

B, side view of the hypopharynx (*Hphy*) with terminal parts of right maxilla (*Mx*) and labium (*Lb*) attached, showing the silk press (*Pr*), its muscles, and the ducts (*Dct*) opening into it, and the spinneret (*Spt*) through which the silk is discharged from the press

C, upper view of the silk press (*Pr*), showing the four sets of muscles (*Mcls*) inserted on its walls and on the rod-like raphe (*Rph*) in its roof

D, side view of the silk press, spinneret, raphe, and muscles

E, cross-section of the silk press, showing its cavity, or lumen (*Lum*), which is expanded by the contraction of the muscles

[287]

from the two glands unite in a median thick-walled sac known as the *silk press* (*Pr*), which opens to the exterior through the spinneret. Two small accessory glands, which look like bunches of grapes and which are sometimes called the *glands of Filippi* (Fig. 155 A, B, C, *GlF*), open into the silk ducts near their front ends.

The relation of the silk ducts and the silk press to the spinneret is seen in the side view of the terminal parts of the labium and the left maxilla, given at B of Figure 155. The silk press (*Pr*) is apparently an organ for regulating the flow of the liquid silk material into the spinneret. It has been supposed, too, that it gives form and thickness to the thread, but the liquid material has still to pass through the rigid tube of the spinneret.

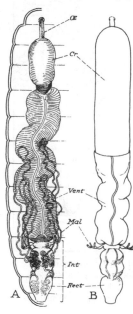

FIG. 156. The alimentary canal of the tent caterpillar

A, before feeding. B, after feeding. *Cr*, crop; *Int*, intestine; *Mal*, Malpighian tubules; *OE*, oesophagus; *Rect*, rectum; *Vent*, ventriculus

The cut end of the press, given at E of Figure 155, shows the crescent form of the cavity (*Lum*) in cross-section, and the thickening in its roof (*Rph*), called the *raphe*. Muscles (*Mcls*) inserted on the raphe and on the sides of the press serve to enlarge the cavity of the press by lifting the infolded roof. The four sets of these muscles in the tent caterpillar are shown at C. The dilation of the press sucks the liquid silk into the cavity through the ducts from the reservoirs, and when the muscles relax, the elastic roof springs back and exerts a pressure on the silk material, which forces the latter through the tube of the spinneret. The continuous passageway from the ducts through the press and into the spinneret is seen from the side at **D.**

The silk liquid is gummy and adheres tightly to whatever it touches, while at the same time it hardens rapidly and becomes a tough, inelastic thread as it is drawn out of the spinneret when the caterpillar swings its head away from the point of attachment.

The mouth of the caterpillar lies between the jaws and the lips. It opens into a short gullet, or *oesophagus*, which, with the pharynx, constitutes the first part of the alimentary canal (Fig. 154, *Phy*, *Oe*). The rest of the canal is a wide tube occupying most of the space within the caterpillar's body and is divided into the *crop* (*Cr*), the stomach, or *ventriculus* (*Vent*), and the *intestine* (*Int*). The crop is a sac for receiving the food and varies in size according to the amount of food it contains (Fig. 156 **A**, **B**, *Cr*). The stomach (*Vent*) is the largest part of the canal. Its walls are loose and wrinkled when it is empty, or smooth and tense when it is full. The intestine (*Int*) consists of three divisions, a short part just back of the stomach, a larger middle part, and a saclike end part called the *rectum* (*Rect*). Six long tubes (*Mal*) are wrapped in many coils about the intestine and run forward and back in long loops over the rear half of the stomach. The three on each side unite into a short basal tube, which opens into the first part of the intestine. The terminal parts of the tubes are coiled inside the muscular coat of the rectum. These tubes are the *Malpighian tubules.*

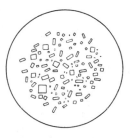

Fig. 157. Crystals from the Malpighian tubules of the tent caterpillar, which are ejected into the walls of the cocoon

When a tent caterpillar goes out to feed, the fore part of its body is soft and flabby; when it returns to the tent the same part is tight and firm. This is because the tent caterpillar carries its dinner home in its crop, digests it slowly while in the tent, and then goes out for more when the crop is empty. It is quite easy to tell by feeling one of these caterpillars whether it is hungry or not. The empty, contracted crop is a small bag contained in the first three segments of the body (Fig. 156 A, *Cr*); but the full crop stretches out to a long cylinder like a sausage, filling the first six segments of the body (B, *Cr*), its rear end sunken into the stomach, and its front end pressed against the back of the head.

The fresh food in the crop consists of a soft, pulpy mass of leaf fragments. As this is passed into the stomach, the crop contracts and the stomach expands, and the caterpillar's center of gravity is shifted backward with the food burden. As the stomach becomes empty there accumulates in it a dark-brown liquid, and it becomes inflated with bubbles of gas. When the caterpillar goes to its meals both crop and stomach are sometimes empty, but usually the stomach still contains some food besides an abundance of the brown liquid and numerous gas bubbles. The refuse that accumulates in the middle section of the intestine is subjected to pressure by the muscles of the intestinal wall, and is here molded into a pellet which retains the imprint of the constrictions and pouches of this part of the intestine and looks like a small mulberry when passed on into the rectum and finally extruded from the body.

The alimentary canal is a tube made of a single layer of cells extending through the body; but its outer surface, that toward the body cavity, is covered by a muscle layer of lengthwise and crosswise fibers, which cause the movement of the food through the canal. The gullet and crop and the intestine are lined internally with a thin cuticula continuous with that covering the surface of the body, and these linings are shed with the body cuticula every time the caterpillar molts.

The Malpighian tubules (Figs. 154, 156 A, *Mal*), being the kidneys of insects, are excretory organs that remove from the blood the waste products containing nitrogen, and discharge them into the intestine along with the waste parts of the food from the stomach. Ordinarily the Malpighian tubules are of a whitish color, but just before the tent caterpillar is ready to spin its cocoon they become congested with a bright yellow substance. Under the microscope this is seen to consist of masses of square, oblong, and rod-shaped crystals (Fig. 157). At this time the caterpillar has ceased to feed and the alimentary canal contains no food or food refuse. The intestine, however,

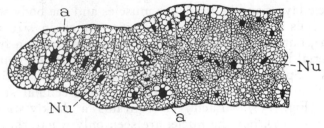

FIG. 158. A piece of the fat-body of the fall webworm
a, a, globules of fatty oil in the cells; *Nu, Nu*, nuclei of the cells

becomes filled with the yellow mass from the Malpighian tubules; and this is the material with which the tent caterpillar plasters the walls of its cocoon, giving them their yellowish color and stiffened texture. The yellow powder of the cocoon, therefore, consists of the crystals from the Malpighian tubules.

We now come to the question of why the caterpillar eats so much. It is almost equivalent to asking, "Why is a caterpillar?" The caterpillar is the principal feeding stage in the insect's life; eating is its business, its reason for being a caterpillar. It eats not only to build up its own organs, many of which are to be broken down to furnish building material for those of the moth, but it eats also to store up within its body certain materials in excess of its own needs, which likewise will contribute to the growth of the moth.

The most abundant of the food reserves stored by the caterpillar is fat. With insects, however, fat does not accumulate among the muscles and beneath the skin. Insects do not become "fat" in external appearance. Their fatty products are held in a special organ called the *fat-body*.

The fat tissue of a caterpillar consists of many small, flat, irregular masses of fat-containing cells scattered all through the body cavity, some of the masses adhering in chains and sheets forming a loose open network about the alimentary canal, others being distributed against the muscle layers and between the muscles and the body wall. The cells composing the tissue vary much in size and shape, but they are always closely adherent, and in fresh material it is often difficult to distinguish the cell boundaries. Specimens prepared and stained for microscopic examination, however, show distinctly the cellular structure (Fig. 158). Each cell contains a darkly-stained nucleus (*Nu*), but the nuclei are seen only where they lie in the plane of the section. The protoplasmic area about the nucleus in each cell appears to be occupied mostly with hollow cavities of various sizes (*a*), but in life each cavity contains a small globule of fatty oil. The protoplasmic material between the oil globules contains also glycogen, or animal starch, as can be shown by staining with iodine. Both fat and glycogen are energy-forming compounds, and their presence in the fat cells of the caterpillar shows that the fat-body serves as a storage organ for these materials during the larval life. The stored fat and glycogen will be consumed during the period of metamorphosis, when the insect is deprived of the power of feeding and receives no further nourishment from the alimentary canal. The transformation processes will then depend upon the food materials that the caterpillar has stored in its own body; and the success of the pupal metamorphosis will depend in large measure on the quantity of these food reserves. A starved caterpillar, therefore, is likely to be unable to accomplish its transformation, or it will produce a dwarfed or an imperfect adult.

How the Caterpillar Becomes a Moth

A short time before the caterpillar is ready to spin its cocoon, it ceases feeding. Its body, as we have just learned, contains now an abundance of energy-giving substances stored in the cells of its fat tissue. When the work of constructing the cocoon is started, the alimentary canal is devoid of food material, the crop is contracted to a narrow cylinder, and the stomach is shrunken and flabby. The stomach, however, contains a mass of soft, orange-brown substance which, when examined under the microscope, is found to consist, not of plant tissue, but of animal cells; it is, in fact, the cellular lining of the caterpillar's stomach which has already been cast off into the cavity of the stomach. The latter is now provided with a new cell wall. The shedding of the old stomach wall marks the first stage in the dismantling of the caterpillar; it is the beginning of the pupal metamorphosis which will convert the caterpillar into the moth. The new stomach wall will first digest and absorb the débris of the old, in order to conserve its proteid materials for the constructive work of the pupa, and it will then itself become transformed into the stomach of the moth.

After the caterpillar has shut itself into the cocoon, its life as a caterpillar is almost ended. Its external appearance is already much altered by the contraction of the body and the loss of the hairy covering, and during the next three or four days a further characteristic change of form takes place. As the body continues to shorten, the first three segments become crowded together; but the abdomen swells out, while the abdominal legs are retracted until they all but disappear. The creature is now (Fig. 159 B) only half the length of the active caterpillar (A), and it would scarcely be recognized as the same individual that so recently spun itself into the cocoon.

During the progress of change in the external form, the caterpillar gradually loses the power of movement. The resultant inactive period in an insect's life, immediately preceding the visible change to the pupa, is called the

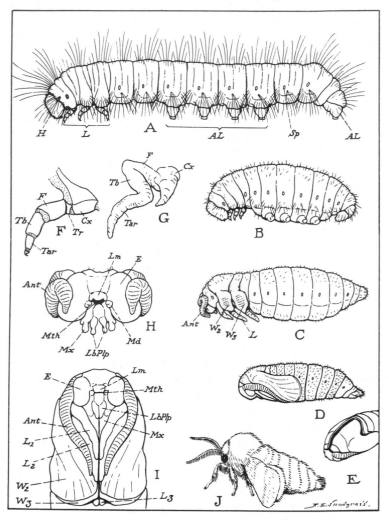

FIG. 159. Transformation of the tent caterpillar into the moth

prepupal stage of the larva. The insect in the prepupal stage has suffered no change in external structure, it still wears the larval skin, and its visible difference from the active larva is a mere alteration in form. Internally, however, important reconstructive processes are now taking place.

The internal activities of reconstruction, which bring about the pupal metamorphosis of the larva to the adult, begin at the head end of the insect and progress posteriorly. They are preceded by a loosening and subsequent detachment of the larval cuticula from the cellular layer of the skin, or *epidermis*, beneath it. The latter, known also as the *hypodermis*, freed now from restraint, enters a period of rapid growth. On the head, the head walls are remodeled and take on a new form, and new antennae and new mouth parts are produced. The new structures have no regard for the forms of the old, though each is produced from a part of the corresponding larval organ. The new antennae, for example, are formed from the larval antennae, but the antennae of the moth are to be much larger than those of the caterpillar. Only the tip, therefore, of each new organ can be formed within the cuticular sheath of the old; the base pushes inward, and the elongating shaft folds against the face of the newly forming head. The same thing is true of the maxillae and labium, but in the case of the mandibles the procedure is simpler, for the jaws are to be reduced in the moth. The epidermal core of each mandible, therefore, simply shrinks within the cuticular sheath of the larval organ, leaving the cavity of the latter almost empty.

As the separation of the cuticula from the epidermis progresses over the region of the thorax and a free space is created between the two layers, the wing buds, which heretofore have been turned inside the caterpillar's body, now evert and come to be external appendages of the pupal body though still covered by the cuticula of the larva (Fig. 159 C, W_2, W_3). The legs of the moth pupa are

INSECTS

formed in the same way as are the antennae and the mouth parts, that is, they are developed from the epidermis of the corresponding larval legs; but, by reason of their increased size, they are forced to bend upward against the sides of the body of the pupa, and, when fully formed, each is found to have only its terminal part within the cuticular sheath of the leg of the caterpillar.

From the thorax, the loosening of the cuticula spreads backward over the abdomen, until at last the entire insect lies free within the cuticular skin of the caterpillar. The so-called prepupal period of the caterpillar, therefore, is scarcely to be regarded as a truly larval stage of the insect. It is still clothed in the larval cuticula, and retains externally all the structural characters of the larva; but the creature itself is in a first growth period of the pupal stage, and may appropriately be designated a *propupa*.

When the cuticula is separated from the epidermis all over the body, it may be cut open and taken off without injury to the wearer. The latter, now a propupa (Fig. 159 C), is then discovered to be a thing entirely different in appearance from the caterpillar. It has a small head bent downward, a thoracic region of three segments, and a large abdomen. The head bears the mouth parts and a pair of large antennae (*Ant*); the thorax carries the wings (W_2, W_3) and the legs (*L*), which latter are much longer than those of the caterpillar, but, being folded beneath the wings, only their ends are visible in side view. The abdomen consists of ten segments and has lost all vestiges of the abdominal legs of the caterpillar (A, *AL*).

Many important changes have taken place in the form and structure of the head and in the appendages about the mouth during the change from the caterpillar to the propupa, as may be seen by comparing Figure 159 H, with Figure 152. Most of the lateral areas of the caterpillar's head (Fig. 152), including the region of the six small eyes on each side, have been converted into the two huge eye areas of the pupa (Fig. 159 H, *E*), which cover the developing compound eyes of the adult. The antennae (*Ant*), as already noted, have increased greatly in size, and they show evidence of their future multiple segmentation. The upper lip, or labrum (*Lm*), on the other hand, is much smaller in the propupa than in the caterpillar, and the great biting jaws of the caterpillar are reduced to mere rudiments in the propupa (*Md*), while the spinneret (Fig. 152, *Spt*) is gone entirely. The labium and the two maxillae are longer and more distinct from each other in the propupa (Fig. 159 H, *Lb*, *Mx*) than in the caterpillar, and their parts are somewhat more simplified. The labium bears two prominent palpi (*LbPlp*).

The remodeling in the external form of the insect proceeds from particular groups of cells in the epidermis, cells that have remained inactive since the time of the embryo, and which, as a consequence, retain an unused vitality. These groups of regenerative epidermal cells, which are the histoblasts, or imaginal discs, of the body wall, have not been particularly studied in the caterpillar; but in certain other insects they have been found to occur in each segment, typically a pair of them on each side of the back, and a pair on each side of the ventral surface. At the beginning of metamorphosis, as the larval cuticula separates from the epidermis, the cells of the discs multiply and spread from their several centers, and the areas newly formed by them take on the contour and structure of the pupa instead of that of the larva. The old cells of the larval epidermis, which have reached the limits of their growing powers and are now in a state of senescence, give way before the advancing ranks of invading cells; their tissues go into dissolution and are absorbed into the body. The new epidermal areas finally meet and unite, and together constitute the body wall of the pupa.

While the new epidermis is giving external form to the pupa beneath the larval cuticula, its cells are generating a new pupal cuticula. As long as the latter is soft and plastic the cell growth may proceed, but when the cuticular substance begins to harden, growth ceases, and the external form of the insect will henceforth show no further change in its structural features.

The propupa of the moth remains for several days a soft-skinned creature (Fig. 159 C) inside the cuticula of the larva, during which time its body contracts in size and its wings, legs, antennae, and maxillæ lengthen. The wings are flattened against the sides of the body, and the other appendages are applied close to the under surface. Then a gluelike substance is exuded from the body wall, which fixes the members in their positions and soon dries into a hard coating or glaze over the body and appendages, giving to the whole a shell-like covering. In this way the soft propupa (C) becomes a chrysalis (D). Finally, the old caterpillar skin splits along the back of the first two body segments, over the top of the head, and down the right side of the facial triangle. The pupa now quickly wriggles out of the enclosing skin and pushes the latter over the rear extremity of its body into the end of the cocoon, where it remains as a shriveled mass, the last evidence of the caterpillar.

The pupa, or chrysalis, of the tent caterpillar (Fig. 159 D) is much smaller than the propupa (C), and its length is only about one-third that of the original caterpillar (A). The color of the chrysalis is at first bright green on the fore parts, yellowish on the abdomen, and usually more or less brown on the back. Soon, however, the color darkens until the front parts and the wings are purplish black, and the abdomen purplish brown. Though the covering of the chrysalis is hard and rigid, the creature is still capable of a very active wriggling of the abdomen, for three of its intersegmental rings remain flexible. By this provision the pupa is able to divest itself of the larval skin. The pupæ of some species of moths push themselves partly out of the cocoon just before the time of transformation to the moth, and when the latter emerges it leaves the pupal skin projecting from the mouth of the cocoon (Plate 12).

Concurrent with the remodeling in the external form of the insect, other changes have been taking place within the body. The first of the complicated metamorphic processes that affect the inner organs occurs in the stomach, where, as we have already observed, the inner wall is cast off at about the time that the caterpillar begins the spinning of its cocoon. This shedding of the stomach lining is quite a different thing from the molting of an external cuticula, for the stomach wall is a cellular tissue. Furthermore, wherever other cell layers are discarded, as in the case of the epidermis, the cells are absorbed *into the body cavity*. A new stomach wall is generated usually from groups of small cells that originally lay outside the old wall and were retained when the latter was cast off. These cells, as do the imaginal discs of the epidermis, form a new lining to the stomach and give a new shape to this organ, which in the adult insect may be quite different from that of the larva. The shedding of the stomach wall is not necessarily a part of the metamorphosis, for in some insects and in certain other related animals, it is said, the stomach epithelium as well as the cuticular lining is shed and renewed with each molt of the body wall.

The parts of the alimentary canal that lie before and behind the stomach, that is, the oesophagus and crop (Fig. 154, *Oe*, *Cr*) and the intestine (*Int*), formed in the embryo as ingrowths of the body wall, are regenerated from groups of cells in their walls in the same manner as is the epidermis itself, the old cells being absorbed into the body. The cuticular linings of these parts are shed with the cuticula of the body wall at the time of the molt. The complete alimentary canal of the moth is very different from that of the caterpillar, as will be shown in the next section of this chapter (Fig. 164).

The walls of the Malpighian tubules are said to be regenerated in some insects, but the tubules do not change

much in form in the moth, and they continue their excretory function during the pupal stage. The silk glands
of the caterpillar are greatly reduced in size, and their
ducts, as a consequence of the suppression of the spinneret,
open at the base of the labium within the entrance to the
mouth.

Internal organs that have not been specially modified
in their development for the purposes of the larva, including usually the nervous system, the heart, the respiratory tubes, and the reproductive organs, suffer little if any
disintegration in their tissues; they simply grow to the
mature form, which may be much more elaborate than
that of the larva, by a resumption of the ordinary processes
of development. The nervous system, and particularly
the tracheal system, however, in some insects undergo
much reconstruction between the larval and the adult
stages.

A most important part of the reconstruction between
the larva and the adult has to do with the muscle system.
Since, in its two active stages, the insect leads usually
two very different lives, the mechanism of locomotion is
likely to be radically different in the larva and in the
adult; and in such cases the transformation of the insect
will involve particularly a thorough reorganization of the
musculature. Most larvae have acquired an elaborate
system of special muscles for their own use because they
have adopted a wormlike mode of progression. On the
other hand, the adults have need of certain muscles, particularly those of the wings, which would be only an en-
cumbrance to a larva. Consequently, muscles needed only
by the adult are suppressed in the larval stage, and the
special muscles of the larva must be cleared away during
the pupal stage. The metamorphosis in the muscle system, therefore, varies much in different insects according
to the mechanical difference between the larva and the
adult.

The purely larval muscles that are to be discarded when

their purpose has been accomplished go into a state of dis-solution during the pupal period. The débris of their tissues is thrown into the blood, from which it is later ab-sorbed as nutriment by the newly forming organs. The caterpillar has a very elaborate system of muscles forming a complicated network of fibers against the inner surface of the body wall, some running longitudinally, others transversely, and still others obliquely. Most of the transverse and oblique fibers are not retained in the moth, and if specimens of those muscles are examined during the early part of the pupal period they are seen to have a weak and abnormal appearance; the structure typical of healthy muscle tissue is obscure or indistinctly evident in them, and in places they are covered with groups of free oval cells. These cells are probably *phagocytes*.

A phagocyte is a blood corpuscle that destroys foreign proteid bodies in the blood, or any unhealthy tissue of the body. It is not probable that the insect phagocytes are the active cause of the destruction of the larval tissues, but they do engulf and digest particles of the degenerating tissues. They are present in large numbers in some insects during metamorphosis, and are scarce or lacking in others. The decadent state of the larval tissues that have passed their period of activity lays them open to the attack of the phagocytes, but these tissues will go into dissolution by the solvent powers of the blood alone. Active, healthy tissues are always immune from phagocytes.

Some of the larval muscles may go over intact to the adult stage, and others may require only a remodeling or an addition of fibers to make them serviceable for the purposes of the adult. The adult muscles that are com-pletely suppressed during the larval stage appear to be generated anew during the pupal stage. There is a dif-ference of opinion among investigators as to how the new muscles are developed, but it is probable that they take their origin from the same tissues that built up the larval muscles.

The development of the internal organs proceeds without interruption from the beginning of the propupal period until the adult organs are completed at the end of the pupal stage. The external parts, however, do not make a continuous growth. After reaching a certain stage of development, the form of the body wall and of the appendages is fixed by the hardening of the new cuticula on their outer surfaces. In this stage, therefore, they must remain, and the half-mature form attained is that characteristic of the pupa. The final development of the body wall and the appendages of the adult is accomplished by a second separation of the epidermis from the cuticula, which allows the cellular layers, now protected by the pupal cuticula, to go through a second period of growth during the pupal stage. This pupal period of growth at last results in the perfection of the external characters of the adult, which are in turn fixed by the formation of the adult cuticula. In the meantime, the new muscles that are to be retained have become anchored at their ends into the new cuticula, and the mechanism of the adult insect is ready for action. The perfect insect, cramped within the pupal shell, has now only to await the proper time for its emergence.

Through the whole period of metamorphosis, the insect must depend on its internal resources for food materials. Oxygen it can obtain by the usual method, for its respiratory system remains functional; but in the matter of food it is in a state of complete blockade. The pupa has two sources of nourishment: first, the food reserves stored in the cells of the fat-body; second, the materials resulting from the breaking down of the larval tissues, which are scattered in the blood and eventually absorbed.

The fat cells, at the beginning of metamorphosis in some insects, give up most of their stored fat and glycogen; and they now become filled with small granules of proteid matter. The proteid granules are probably elaborated in the fat cells from the absorbed detritus of the larval organs by means of enzymes produced in the nuclei of the cells. The fat cells thus take on the function of a stomach, converting the materials dissolved in the blood into forms that the growing tissues can assimilate. During this time the masses of fat tissue that compose the fat-body of the

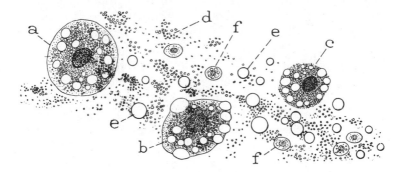

FIG. 160. Bodies in the blood of a young pupa of the tent caterpillar

a, a free fat cell, containing large oily fat globules, and small proteid granules; *b*, *c*, fat cells in dissolution; *d*, free proteid granules in the blood, and *e*, fat globules liberated from the disintegrating fat cells; *f*, blood corpuscles

larva have broken up into free cells, and these cells, vacuolated with oil globules and later charged with proteid granules, now fill the blood.

The interior of the moth pupa, or chrysalis, shortly after the larval skin is shed, contains a thick, yellow, creamy liquid. In it there may be discovered, however, the alimentary tract, the nervous system, and the tracheal tubes, the latter filled with air; but all these parts are so soft and delicate that they can scarcely be studied by ordinary methods of dissection.

The creamy pulp of the pupa's body, when examined under the microscope, is seen to consist of a clear, pale, amber-yellowish liquid full of small bodies of various sizes (Fig. 160), which give it the opaque appearance and thick consistency. The liquid medium is the blood, or body lymph. The largest bodies in it are free fat cells (*a*); smaller ones are probably blood corpuscles (*f*); and the finest matter consists of great quantities of minute grains (*d*) floating about separately or adhering in irregular masses. Besides these elements there are many droplets of oil (*e*), recognizable by their smooth spherical outlines and golden-brown color. The fat cells are mostly irregularly ovoid or elliptical in shape; their protoplasm is filled with large and small oil globules, and contains also masses of fine granules like those floating free in the blood. These granules are the protoplasmic substances formed within the fat cells. Many of the cells have irregular or broken outlines (*b*, *c*), as if their outer walls had been partly dissolved, and the contents of such cells appear to be escaping from them. In fact, many are clearly in a state of dissolution, discharging both their oil globules and their proteid inclusions into the blood; and it is clear that the similar matter scattered so profusely through the blood liquid has come from fat cells that have already disintegrated. All these materials will gradually be consumed in the building of the tissues of the adult, the organs of which are now in process of formation.

In Chapter IV we learned that every animal consists of a body, or *soma*, formed of cells that are differentiated from the germ cells usually at an early stage of development. The function of the soma is to give the germ cells the best chance of accomplishing their purpose. An insect that goes through two active forms during its life, a larval and an adult form, differs from other animals in having a *double soma*. The entire organism, of course, is not double, for, as we have just seen in the study of the caterpillar, many of the more vital organs are continuous from the larva to the adult; but there is a group of organs which, after reaching a definite form of development in the larval stage, at the end of this stage virtually die and go into dissolution, while a new set of tissues develops into new organs or into new tissues replacing those that have been lost. The groups of somatic cells that form the tissues and organs that undergo a metamorphosis, therefore, are differentiated in the embryo into two sets of cells, one set of which will form the special organs of the larva, while those of the other will remain dormant during the larval life to form the adult organs when the larval cells have completed their functions. The cells of the second set carry the hereditary influences that will cause them to develop into the original, or ancestral, form of the species; the cells of the first set produce the temporary larval form, which may retain certain primitive characters from the embryonic stage, but which does not represent an ancestral form in the evolution of the species.

An extreme case of anything is always more easily understood when we can trace it back to something simple, or link it up with something familiar. The metamorphosis of insects appears to be one of the great mysteries of nature, but reduced to its simplest terms it becomes only an exaggerated case of a temporary growth in certain groups of cells to form something of use to the young, which disappears by resorption when the occasion for its use is past. Innumerable simple cases of this kind might be cited from insects; but there is a familiar case of well-developed metamorphosis even in our own growth, namely, the temporary development of the milk teeth and their later substitution by the adult teeth. If a similar process of double growth from the somatic cells had been carried to other organs, we ourselves should have a metamorphosis entirely comparable with that of insects.

THE MOTH

For three weeks or a little longer the processes of reconstruction go on within the pupa of the tent caterpillar, and then the creature that was a caterpillar breaks through its coverings and appears in the form and costume of a moth (Fig. 159 J). The pupal shell splits open on the forward part of the back (E) to allow the moth to emerge, but the latter then only finds itself face to face with the wall of the cocoon. It has left behind its cutting instruments, the mandibles, with its discarded overalls; but it has turned chemist and needs no tools. The glands that furnished the silk for the larva have shrunken in size and have taken on a new function; they now secrete a clear liquid that oozes out of the mouth of the moth and acts as a solvent on the adhesive surfaces of the cocoon threads. The strands thus moistened are soon loosened from one another sufficiently to allow the moth to poke its head through the cocoon wall and force a hole large enough to permit of its escape. The liquid from the mouth of the moth turns the silk of the cocoon brown, and the lips of the emergence hole are always stained the same color— evidence that it is this liquid that softens the silk—and the frayed edges of the hole left in the cocoon of the tent caterpillar show many loose ends of threads broken by the moth in its exit.

The most conspicuous features of the moth (Fig. 161) are its furry covering of hairlike scales and its wings. The wings are short when the insect first emerges from its cocoon (Fig. 159 J), but they quickly expand to normal length and are then folded over the back (Fig. 161 A). The colors of the moths of the tent caterpillar are various shades of reddish-brown with two pale bands obliquely crossing the wings (Plate 14 G, H). The female moth (Plate 14 H, Fig. 161 B) is somewhat larger than the male, her body being a little over three-fourths of an inch in length, and the expanded wings one and three-fourths inches across.

The tent caterpillars perform so thoroughly their duty of eating that the moths have little need of more food. Consequently the moths are not encumbered with implements of feeding. The mandibles, which were such large and important organs in the caterpillar (Fig. 152, *Md*) but which shrank to a rudimentary condition in the propupa (Fig. 159 H, *Md*), are gone entirely in the moth (Fig. 162). The maxillae, which were fairly long lobes in the propupa (Fig. 159 H, *Mx*), have likewise been

[306]

reduced to mere rudiments in the adult, where they appear as two insignificant though movable knobs (Fig. 162, *Mx*). The median part of the labium has been reduced to almost nothing in the moth; but the labial palpi (*LbPlp*) are long and three-segmented, and when normally covered with hairlike scales they form two conspicuous feathery brushes that project in front of the face.

The mouth parts of the tent caterpillar moth are not typical of these organs of moths and butterflies in gen-

FIG. 161. Moths of the tent caterpillar, *Malacosoma americana*. (A little greater than natural size)

eral, for most of these insects are provided with a long *proboscis* by means of which they are able to feed on liquids. Everyone is familiar with the large humming-bird moths, or hawk moths, that are to be seen on summer evenings as they dart from flower to flower, thrusting into each corolla a long tube uncoiled from beneath the head; and we have all seen the sunlight-loving butterflies carelessly flitting over the flower beds, alighting here and there on attractive blooms to sip the sweet liquid from the nectar cups.

Moths and butterflies carry the proboscis tightly coiled, like a tiny watch spring (Fig. 163 A, *Prb*), beneath the head and just behind the mouth. It can be unwound and extended (B, *Prb*) whenever the insect wants to extract a drop of nectar from the depths of a

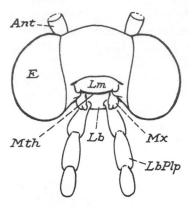

FIG. 162. Facial view of the head of the tent caterpillar moth, with covering scales removed, and antennae cut off near their bases

Ant, base of antenna; *E*, compound eye; *Lb*, labium; *LbPlp*, labial palpus; *Lm*, labrum; *Mth*, mouth; *Mx*, maxilla

flower corolla, or when it would merely take a drink of water or other liquid. The proboscis consists of the greatly lengthened maxillae firmly attached to each other by dovetailed grooves and ridges. The inner face of each maxilla is hollowed in the form of a groove running its entire length, and the two grooves apposed between the united maxillae are converted into a central channel of the proboscis. The two blades of the proboscis spring from the sides of the mouth. The first part of the alimentary canal just back of the mouth is transformed into a bulblike sucking apparatus. The

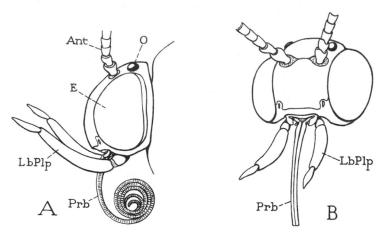

FIG. 163. Head and mouth parts of the peach borer moth

A, side view. B, three-quarter facial view. *Ant*, basal part of antenna; *E*, compound eye; *LbPlp*, labial palpus; *O*, ocellus; *Prb*, proboscis

[308]

upper wall of the structure is ordinarily collapsed into the cavity of the bulb, but it is capable of being lifted by strong muscles inserted upon it from the walls of the head. The alternate opening and closing of the bulb sucks the liquid food up through the tube of the proboscis and forces it back into the gullet. The moths and butterflies are thus sucking insects, as are the aphids and cicadas, but they are not provided with piercing organs, though some species have a rasp at the end of the proboscis which is said to enable them to obtain juices from soft-skinned fruits.

With the tent caterpillar, it is interesting to note, the maxillae are much longer in the pupa (Fig. 159 I, Mx) than they are in either the caterpillar or the adult moth (Fig. 162, Mx), as if nature had intended the tent caterpillar moth to have a proboscis like that of other moths, but had then changed her mind. The real meaning of this is that the moths of the present-day tent caterpillars are descended from ancestors that had a functional proboscis in the adult stage like that of other moths, and that the reduction of the proboscis of the modern moths has taken place in times so recent that the organ has not yet been suppressed to the same degree in the pupa.

The alimentary tract of the tent caterpillar moth is very different from that of the caterpillar. In the caterpillar, the organ consists of three principal parts (Fig. 164 A), the first comprising the oesophagus (Oe) and the crop (Cr), the second being the stomach, or ventriculus ($Vent$), and the third the intestine (Int). In an adult moth that is almost mature, but which is still inside the pupal shell (B), the oesophagus has become a long narrow tube (Oe) at the rear end of which the crop forms a small sac (Cr) projecting upward, which may contain a bubble of gas. The stomach has contracted to a pear-shaped bag with very thin transparent walls, and is usually filled with a dark-brown liquid. The intestine has changed radically in form, for it now consists of a long, slender, tubular part, the small intestine (*SInt*), and of a great terminal receptacle, the rectum (*Rect*), filled with a mass of soft orange-colored matter. In the fully-matured insect (C), after it has escaped from the cocoon, still further

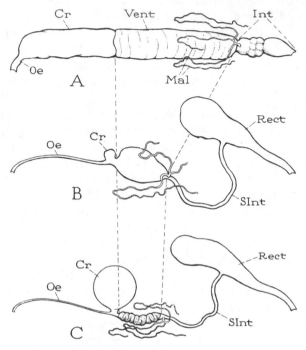

Fɪɢ. 164. Transformation in the form of the alimentary canal of the tent caterpillar from the larva to the adult moth

A, alimentary canal of the caterpillar. B, the same of the pupa. C, the same of the moth

Cr, crop; *Int*, intestine; *Mal*, Malpighian tubules (not shown full length); *Oe*, oesophagus; *Rect*, rectum; *SInt*, small intestine; *Vent*, ventriculus

alterations have taken place. The crop sac (*Cr*) is now greatly distended into a spherical vesicle tensely filled with gas—air, probably, that the moth has swallowed, perhaps to aid it in breaking the pupal shell, for there are sometimes small bubbles also in the tubular oesophagus.

The stomach is contracted to a mere remnant of its former size (A, *Vent*), and its walls are thrown into thick corrugations. The intestine (*SInt*) is about the same as in the earlier stage of the moth (B).

Since the moth of the tent caterpillar probably eats nothing, it has little use for a stomach. The intestine, however, must serve as an outlet for the Malpighian tubules (*Mal*), since the latter remain functional through the pupal stage. The secretion of the tubules contains great numbers of minute spherical crystals, which accumulate in the rectal sac (*Rect*) where they form the orange-colored mass contained in this organ and discharged as soon as the moth leaves the cocoon.

Most of the male moths of the tent caterpillar emerge from the cocoons several days in advance of the females. At this time their bodies contain an abundance of fat which fills the cells of the fat tissue as droplets of oil. This fat is probably an energy-forming reserve which the male moth inherits from the caterpillar, for the internal reproductive organs are not yet fully developed and do not become functional until about the time the females are out of their cocoons.

The bodies of the female tent caterpillar moths, on the other hand, contain little or no fat tissue; but each female is fully matured when she emerges from the cocoon, and her ovaries are full of ripe eggs ready to be laid as soon as the fertilizing element is received from the male (Fig. 165, *Ov*). The spermatozoa will be stored in a special receptacle, the *spermatheca* (*Spm*), which is connected with the exit duct of the ovaries (*Vg*) by a short tube. Each egg is then fertilized as it issues from the oviduct. The material that will form the covering of the eggs when laid is a clear, brown liquid contained in two great sacs (Fig. 165, *Res*) that open into the end of the median oviduct (*Vg*). Each sac is the reservoir of a long tubular gland (*ClGl*). The liquid must be somehow mixed with air when it is discharged over the eggs to give the egg covering its frothy texture. It soon sets into a jellylike substance, then becomes firm and elastic like soft rubber, and finally turns dry and brittle.

The date of the egg laying depends on the latitude of the region the moths inhabit, varying from the middle

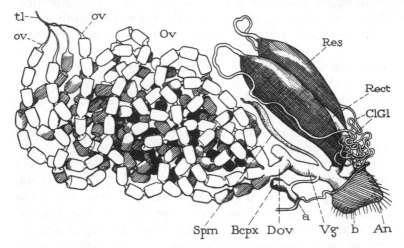

FIG. 165. The female reproductive organs of the moth of the tent caterpillar, as seen from the left side

a, external opening of the bursa copulatrix; *An*, anus; *b*, opening of the vagina; *Bcpx*, bursa copulatrix; *ClGl*, colleterial glands, which form the substance of the egg covering; *Dov*, duct of the left ovary; *Ov*, left ovary, full of ripe eggs; *ov, ov*, upper ends of the ovarian tubules; *Rect*, rectum; *Res*, reservoirs of the colleterial glands (*ClGl*); *Spm*, sperma theca, a sac for the storage of the spermatozoa; *tl*, terminal strand of the ovary; *Vg*, vagina

of May in the southern States to the end of June or later in the north. While the eggs will not hatch until the following spring, they nevertheless begin to develop at once, and within six weeks young caterpillars may be found fully formed within them (Fig. 166 B). Each little caterpillar has its head against the top of the shell and its body bent U-shaped, with the tail end turned a little to one side. The long hairs of the body are all turned forward and form a thin cushion about the poor creature, which for crimes yet uncommitted is sentenced to eight

months' solitary confinement in this most inhuman position. Yet, if artificially liberated, the prisoner takes
no advantage of the freedom offered. Though it can
move a little, it remains coiled (A) and will fold up again
if forcibly straightened, thus asserting that it is more com-
fortable than it looks.

It is surprising that these infant caterpillars can remain
inactive in their eggshells all through the summer, when
the warmth spurs the vitality of other species and speeds
them up to their most rapid growth and development.
External conditions
in general appear to
have much to do with
regulating the lives of
insects, and if the tent
caterpillars in their
eggs seem to give
proof that the crea-
tures are not entirely
the slaves of environ-
ment, the truth is
probably that all in-
sects are not gov-
erned by the same

Fig. 166. The young tent caterpillar fully
formed within the egg by the middle of
summer

A, the young caterpillar removed from the
egg. B, the caterpillar in natural position
within the egg

conditions. We have seen that some of the grasshoppers
and some of the aphids will not complete their development except after being subjected to freezing temperatures, and so it probably is with the tent caterpillars—
it is not warmth, but a period of cold that furnishes the
condition necessary to the final completion of their development. Whatever may be the secret source of their
patience, however, the young tent caterpillars will bide
their time through all the heat of summer, the cold of
winter, and not till the buds of the cherry or apple leaves
are ready to open the following spring will they awake
and gnaw through the inclosing shells against which their
faces have been pressing all this while.

CHAPTER X

MOSQUITOES AND FLIES

THOUGHTFUL persons are much given to pondering on what is to be the outcome of our present age of intensive mechanical development. Thinking, the writer holds, is all right as a means of diverting the mind from other things, but those who make a practice or a profession of it should follow the example of that famous thinker of Rodin's, who has consistently preserved a most commendable silence as to the nature of his thoughts. We can all admire thinking in the abstract; it is the expression of thoughts that disturbs us. So it is that we are troubled when the philosophers warn us that the development of mechanical proficiency is not synonymous with advancement of true civilization. However, it is not for an entomologist to enter into a discussion of such matters, because an observer untrained in the study of human affairs is as likely as not to get the impression that only a very small percentage of the present human population of the world is devoted to efficiency in things mechanical or otherwise.

There is no better piece of advice for general observance than that which admonishes the cobbler to stick to his last, and the maxim certainly implies that the entomologist should confine himself to his insects. However, we can not help but remark how often parallelisms are to be discovered between things in the insect world and affairs in the human world. So, now, when we look to the insects for evidence of the effect of mechanical perfection, we observe with somewhat of a shock that those very insect species which unquestionably have gone farthest along the road of mechanical efficiency have produced little else commendable. In this class we would place the mosquitoes and the flies; and who will say that either mosquitoes or flies have added anything to the comfort or enjoyment of the other creatures of the world?

Reviewing briefly the esthetic contributions of the major groups of insects, we find that the grasshoppers have produced a tribe of musicians; the sucking bugs have evolved the cicada; the beetles have given us the scarab, the glow-worm, and the firefly; the moths and butterflies have enriched the world with elegance and beauty; to the order of the wasps we are indebted for the honeybee. But, as for the flies, they have generated only a great multitude of flies, amongst which are included some of our most obnoxious insect pests.

However, in nature study we do not criticize; we derive our satisfaction from merely knowing things as they are. If our subject is mosquitoes and flies, we look for that which is of interest in the lives and structure of these insects.

FLIES IN GENERAL

The mosquitoes and the flies belong to the same entomological order. That which distinguishes them principally as an order of insects is the possession of only one pair of wings (Fig. 167). Entomologists, for this reason, call the mosquitoes and flies and all related insects the *Diptera*, a word that signifies by its Greek components "two wings." Since nearly all other winged insects have four wings, it is most probable that the ancestors of the winged insects, including the Diptera, had likewise two pairs of wings. The Diptera, therefore, are insects that have become specialized primarily during their evolution by the *loss* of one pair of wings.

We shall now proceed to show that the evolution of a two-winged condition from one of four wings has been a progress toward greater efficiency in the mechanism of flight, and that the acme in this line has been attained by the flies and mosquitoes. The truth of this contention will become apparent when we compare the relative development of the wings and the manner or effectiveness of flight in the several principal orders of insects.

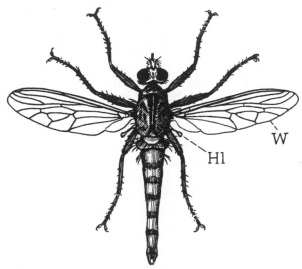

FIG. 167. A robber fly, showing the typical structure of any member of the order Diptera

The flies are two-winged insects, the hind wings being reduced to a pair of knobbed stalks, the halteres (*Hl*)

It is most probable that when insects first acquired wings the two pairs were alike in both size and form. The termites (Fig. 168 A) afford a good example of insects in which the two pairs of wings are still almost identical. Though the termites are poor flyers, their weakness of flight is not necessarily to be attributed to the form of the wings, because their wing muscles are partially degenerate. The dragonflies (Fig. 58) are particularly strong flyers, and with them the two pairs of wings are but little different in size and form; but the dragonflies are provided with sets of highly developed wing muscles which are much more effective than those of other insects. From these examples, therefore, we can not well judge of the mechanical efficiency of two pairs of equal wings moved by the equipment of muscles possessed by most

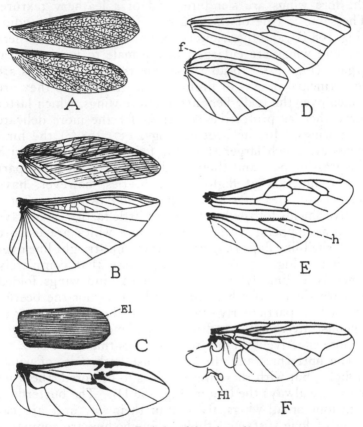

Fig. 168. Evolution of the wings of insects

A, wings of a termite, approximately the same in size and shape.   B, wings of a katydid, the hind wings are the principal organs of flight.   C, wings of a beetle, the fore wings changed to protective shells, elytra (*El*), covering the hind wings.   D, wings of a hawk moth, united by the spine (*f*), which is held in a hook on under surface of fore wing.   E, wings of the honeybee, held together by hooks (*h*) on edge of hind wing.   F, wing of a blowfly, and the rudimentary hind wing, or halter (*Hl*)

insects; but it is evident that the majority of insects have found it more advantageous to have the fore and hind wings different in one way or another.

In the grasshoppers, it was observed (Fig. 63), the hind wings are expanded into broad membranous fans, while the fore wings are slenderer and of a leathery texture. The same is true of the roaches (Fig. 53), the katydids (Fig. 168 B), and the crickets, except in special cases where the fore wings are enlarged in the male to form musical organs (Fig. 39). In all these insects the hind wings are the principal organs of flight. When not in use they are folded over the body beneath the fore wings, which latter serve then as protective coverings for the more delicate hind wings. In the beetles (Figs. 137, 168 C) the hind wings are much larger than the fore wings, and, as with the grasshoppers and their kind, they take the chief part in the function of flight. The beetles, however, have carried the idea of converting the fore wings into protective shields for the hind wings a little farther than have the grasshoppers; with them the fore wings are usually hard, shell-like flaps that fit together in a straight line over the back (Fig. 137 A), forming a case that completely conceals, ordinarily, the membranous hind wings folded beneath them. Neither the grasshoppers nor the beetles are swift or particularly efficient flyers, but they appear to demonstrate that the ordinary insect mechanism of flight is more effective with one pair of wings than with two.

The butterflies and the moths use both pairs of wings in flight; but with these insects, it is to be noted, the *front* wings are always the larger (Fig. 168 D). The butterflies, with four broad wings, fly well in their way and are capable of long-sustained flight, though they are comparatively slow goers. Some of the moths do much better in the matter of speed, but it is found that the faster flying species have the fore wings highly developed at the expense of the hind wings; and that the two wings on each side, furthermore, are yoked together in such a manner as to insure their acting as a single wing (D). The moths clearly show, therefore, as do the grasshoppers and the beetles, the efficiency of a single pair of flight organs as opposed to two. The moths, however, have attacked from a different angle the problem of converting their inherited equipment of four wings into a two-wing mechanism—instead of suppressing the flight function in one pair of wings, they have given a mechanical unity to the two wings of each side, thus attaining functionally a two-winged condition.

The wasps (Fig. 133) and bees, likewise, have evolved a two-winged machine from a four-wing mechanism on the principle of uniting the two wings on each side. The bees have adopted a particularly efficient method of securing the wings to each other, for each hind wing is fastened to the wing in front of it by a series of small hooklets on its anterior vein that grasp a marginal thickening on the rear edge of the front wing (Fig. 168 E). Moreover, the bees have so highly perfected the unity in the design of the wings that only on close inspection is it to be seen that there are actually two wings on each side of the body.

Finally, the flies, including all members of the order Diptera, have boldly executed the master stroke by completely eliminating the second pair of wings from the mechanism of flight. The flies are literally two-winged insects (Figs. 167, 168 F). Remnants of the hind wings, it is true, persist in the form of a pair of small stalks, each with a swelling at the end, projecting from behind the bases of the wings (Figs. 167, 168 F, *Hl*). These stalks are known as "balancers," or *halteres*, and in their structure they preserve certain features that show them to be rudiments of wings.

The giving over of the function of flight to the front pair of wings has necessarily involved a reconstruction in the entire framework and musculature of the thorax, and a study of the fly thorax gives a most interesting and instructive lesson in the possibilities of adaptive evolution, showing how a primitive ancestral mechanism may be entirely remodeled to serve in a new capacity. If the flies had been specially "created," and not evolved, their structure could have been much more directly fitted to their needs.

It is not only in the matter of wings and the method of flight that the flies show they are highly evolved insects;

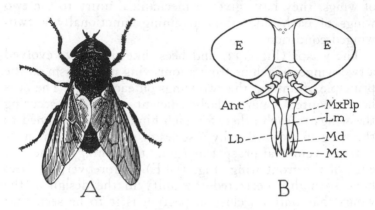

FIG. 169. The black horsefly, *Tabanus atratus*
A, the entire fly. B, facial view of the head and mouth parts. *Ant*, antenna;
E, E, compound eyes; *Lb*, labium; *Lm*, labrum; *Md*, mandible; *Mx*, maxilla;
MxPlp, maxillary palpus

they are equally specialized in the structure of their mouth parts and in their manner of feeding. The flies subsist on liquid food. Those species that can satisfy their wants from liquids freely accessible have the mouth parts formed for sucking only. Unfortunately, however, as we all too well know, there are many species that demand, and usually obtain, the fresh blood of mammals, including that of man, and such species have most efficient organs for piercing the skin of their victims.

The most familiar examples of flies that "bite" are the mosquitoes and horseflies. The horseflies (Fig. 169 A), some of which are called also gadflies and deer flies, belong to the family Tabanidae. An examination of the head of

the common large black horsefly (Fig. 169 B) will show
the nature of the feeding organs with which these flies are
equipped. Projecting downward from the lower part of
the head are a number of appendages; these are the mouth
parts. They correspond in number and in relative position with the mouth appendages of the grasshopper (Fig.
66), but they differ from the latter very much in form
because they are adapted to quite a different manner of
feeding. The horsefly does not truly bite; it pierces the
skin of its victim and sucks up the exuding blood.

By spreading apart the various pieces that compose the
group of mouth parts of the horsefly, it will be seen that
there are nine of them in all. Three are median in position, and therefore single, but the remaining six occur in
duplicate on the two sides, forming thus three sets of
paired structures. The large club-shaped pieces, however, that lie at the sides of the others, are attached at
their bases to the second paired organs and constitute a
part of the latter, so that there are really only two sets of
paired organs. The anteriormost single piece is the
labrum (Fig. 169 B, *Lm*); the first paired organs are the
mandibles (*Md*); the second are the maxillae (*Mx*); the
second median piece is the hypopharynx (not seen in
Fig. 169 B); and the large, unpaired, hindmost organ is the
labium (*Lb*). The lateral club-shaped pieces are the palpi
of the maxillae (*MxPlp*).

The labrum is a strong, broad appendage projecting
downward from the lower edge of the face (Figs. 169 B,
170 A, *Lm*). Its extremity is tapering, but the tip is
blunt; its under surface is traversed by a median groove
extending from the tip to the base but closed normally
by the hypopharynx (Fig. 170 D, *Hphy*), which fits against
the under side of the labrum and converts the groove into
a tube. The upper end of this tube leads directly into the
mouth, a small aperture situated between the base of the
labrum and the base of the hypopharynx and opening into
a large, stiff-walled, bulblike structure (Fig. 170 A, *Pmp*)

which is the mouth cavity. The anterior wall of the bulb is ordinarily collapsed, but it can be lifted by a set of strong muscles (*Mcl*) arising on the front wall of the head (*Clp*). This bulb is the sucking pump of the fly, and it will be

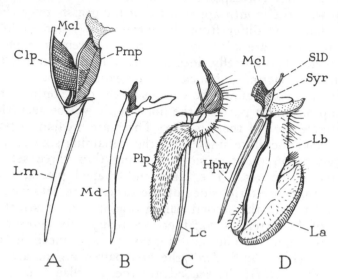

FIG. 170. Mouth parts of a horsefly, *Tabanus atratus*

A, the labrum (*Lm*) and mouth pump (*Pmp*), with dilator muscles of the pump (*Mcl*) arising on the clypeal plate (*Clp*) of the head wall. The mouth is behind the base of the labrum

B, the left mandible

C, the left maxilla, consisting of a long piercing blade (*Lc*), and a large palpus (*Plp*)

D, the labium (*Lb*) terminating in the large labella (*La*), and the hypopharynx (*Hphy*) showing the salivary duct (*SlD*) and its syringe (*Syr*), discharging into a channel of the hypopharynx (*Hphy*) that opens at the tip of the latter

seen that it is very similar to that of the cicada (Fig. 122, *Pmp*). In the fly, however, the liquid food is drawn up to the mouth through the labro-hypopharyngeal tube instead of through a channel between the appressed maxillae.

The mandibles of the horsefly (Fig. 170 B, *Md*) are long, bladelike appendages, very sharp pointed, thickened on the outer edges and thin on the knifelike inner edges. They appear to be cutting organs, for each is articulated to the lower rim of the head by its expanded base in such a manner that it can swing sidewise a little but can not be protruded and retracted as can the corresponding organ of the cicada. The maxillae (C) are slender stylets, each supported on a basal plate attached to the head; this plate carries also the large, two-segmented palpus (*Plp*). The maxillae are probably the principal piercing tools of the horsefly's mouth-part equipment.

The median hypopharynx (Fig. 170 D, *Hphy*) is a tapering blade somewhat hollowed above, normally appressed, as just observed, against the under surface of the labrum to form the floor of the food canal. The hypopharynx itself is traversed by a narrow tube which is a continuation from the salivary duct (*SlD*). The latter, however, just before it enters the base of the hypopharynx, is enlarged to form an injection syringe (*Syr*). The salivary syringe in structure is a small replica of the mouth pump (A, *Pmp*), and its muscles arise on the back of the latter. The saliva of the fly is injected into the wound from the tip of the hypopharynx. By reason of this fact, the bite of a fly may be the source of infection to the victim, for it is evident that the injection of saliva affords a means for the transfer of internal disease parasites from one animal to another.

Behind all the parts thus far described is the median labium (Fig. 170 D, *Lb*), a much larger organ than any of the others, consisting of a thick basal stalk and two great terminal lobes (*La*). The soft, membranous under surfaces of the lobes, which are known as the *labella*, are marked by the dark lines of many parallel, thick-walled grooves extending crosswise. These grooves may be channels for collecting the blood that exudes from the wound, or they may also distribute the saliva as it issues from the tip of the hypopharynx between the ends of the labella. The effect of the saliva of the horsefly on the blood is not known, but the saliva of some flies is said to prevent coagulation of the blood.

Some of the smaller horseflies will give us an unsolicited sample of their biting powers, and in shaded places along roads they often make themselves most vexatious to the foot traveler just when he would like to sit down and enjoy a quiet rest. To horses, cattle, and wild mammals, however, these flies are extremely annoying pests, and, where abundant, they must make the lives of animals almost unendurable; for the sole means of protection the latter have against the painful bites of the flies is a swish of the tail, which only drives the insects to make a fresh attack on some other spot.

There is another family of "biting" flies, known as the robber flies, or Asilidae (Fig. 167), the members of which attack other insects. They are strong flyers and take their victims on the wing, even bees falling prey to them. The robber flies have no mandibles, and the strong, sharp-pointed hypopharynx appears to be the chief piercing implement. The saliva of the fly injected into the wound dissolves the muscles of the victim, and the predigested solution is then completely sucked out.

As was shown in Chapter VIII, on metamorphosis, whenever the adult form of an insect is highly specialized for a particular kind of life, it is usually found that the young is also specialized but in a way of its own to adapt it to a manner of living quite different from that of its parent. This principle is particularly true of the flies, for, if the adult flies are to be regarded as in general the most highly evolved in structure of all the adult insects, there can be no doubt that the young fly is the most highly specialized of all the insect larvae.

The flies belong to that large group of insects which do not have external wings in the larval stage, but with the flies the suppression of the body appendages includes also the legs, so that their larvae are not only wingless but legless as well (Fig. 171). The legs, however, as the wings, are represented by internal buds, which, when they enter the period of growth during the early stage of metamorphosis, are turned inside out to form the legs of the adult fly.

The lack of legs gives a cylindrical simplicity of form to most fly larvae, which not only makes these insects look like worms, but has caused many of them to live the life of

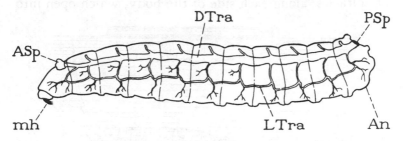

FIG. 171. Structure of a fly larva, or maggot

An, anus; *ASp*, anterior spiracle; *DTra*, dorsal tracheal trunk; *LTra*, lateral tracheal trunks; *mh*, mouth hooks; *PSp*, posterior spiracle

a worm and to adopt the ways of a worm. In compensation for the loss of legs, the fly larvae are provided with an intricate system of muscle fibers lying against the inner surface of the body wall, which enables them to stretch and contract and to make all manner of contortionistic twists.

At first thought it seems remarkable that a soft-bodied, wormlike creature can stretch itself by muscular contraction. It must be remembered, however, that the body of the larva is filled with soft tissues, many of which are but loosely anchored, and that the spaces between the organs are filled with a body liquid. The creature is, therefore, capable of performing movements by making use of its structure as a hydraulic mechanism; a contraction of the rear part of the body, for example, drives the body liquid and the soft movable organs forward, and thus extends the anterior parts of the body. A contraction of the lengthwise muscles then pulls up the rear parts, when the movement of extension may be repeated. In this fashion the soft, legless larva moves forward; or, by a reversal of the process when occasion demands, it goes backward.

A special feature in the construction of fly larvae is the arrangement of their breathing apertures, which is correlated with the manner of breathing. In most insects, as we have learned (Fig. 70), there is a row of breathing pores, or spiracles, along each side of the body, which open into

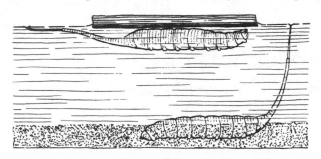

FIG. 172. Rat-tailed maggots, larvae of the drone fly, which live submerged in water or mud and breathe at the surface through a long, tail-like respiratory tube

Upper figure, resting beneath a small floating object; lower, feeding in mud at the bottom

lateral tracheal trunks. In the fly larva, however, these spiracles are closed and are not opened for respiration until the final change of the pupa to the adult.

The fly larva is provided with one or two pairs of special breathing organs situated at the ends of the body. Some species have a pair of these organs at each end of the body (Fig. 171, *ASp*, *PSp*), and some a pair at the posterior end only. The anterior organs, when present (Fig. 171, *ASp*), consist of perforated lobes on the first body segment, the pores of which communicate with the anterior ends of a pair of large dorsal tracheal trunks (*DTra*). The posterior organs (*PSp*) consist of a pair of spiracles on the rear end of the body, which open into the posterior ends of the dorsal tracheae. By means of this respiratory arrangement, the fly larva can live submerged in water, or buried in mud or any other soft medium, so long as it keeps one end of the body out for breathing.

The rat-tailed maggot (Fig. 172), which is the larva of a large fly that looks like a drone bee, has taken a special advantage of its respiratory system; for the rear end of its body, bearing the posterior spiracles, is drawn out into a long, slender tube. The creature, which lives in foul water or in mud, can by this contrivance hide itself beneath a floating object and breathe through its tail, the tip of which may come to the surface of the water at a point some distance away. The end of the tail is provided with a circlet of radiating hairs surrounding the spiracles, which keeps the tip of the tail afloat and prevents the water from entering the breathing apertures.

The great disparity of structure between the larva of a fly and the adult necessarily involves much reconstruction during the period of transformation, and probably the inner processes of metamorphosis are more intensive in the more highly specialized Diptera than in any other group of insects.

The pupa of an insect, as we have seen in Chapter VIII (page 254), is very evidently a preliminary stage of the adult, the larval characters being usually discarded with the last molt of the larva. The pupa of most flies, however, while it has the general structure of the adult fly (Fig. 182 A, F), retains the special respiratory scheme of the larva and at least a part of the larval breathing organs. The fact that the larvae breathe through special spiracles located on the back suggests that the primitive fly larvae lived in water or in soft mud, and that it was through an adaptation to such an environment that the lateral spiracles were closed and the special dorsal spiracles developed. The retention by many fly pupae of the larval method of breathing and of at least a part of the larval respiratory organs, though their habitat would not seem necessarily to demand it, suggests, furthermore, that the pupae of the ancestors of such species lived in the same medium as the larvae.

If our supposition is correct, we may see a reason for the apparent exception in the flies to the general rule that the pupa presents the adult structure and discards the peculiarly larval characters. The pupae of some flies whose

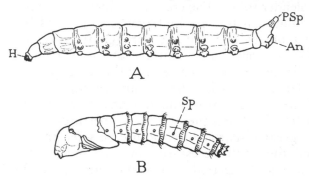

FIG. 173. Larva (A) and pupa (B) of a horsefly, *Tabanus puncti-fer* (about 1½ times natural size)

An, anus; *H*, head; *PSp*, posterior spiracle; *Sp*, spiracle

larvae live in the water, however, revert at once to the adult system of lateral spiracles (Fig. 173 B, *Sp*). With such species, the larva comes out of the water just before pupation time and transforms in some place where breathing is possible by the ordinary respiratory organs. This is the general rule with other insects whose larvae are aquatic.

The order of the Diptera is a large one, and we might go on indefinitely describing interesting things about flies in general. Such a course, however, would soon fill a larger book than this; hence, since we are already in the last chapter, a more practical plan will be to select for special consideration a few species that have become closely associated with the welfare of man or of his domesticated animals. Such species include the mosquitoes, the house fly, the blowfly, the stable fly, the tsetse fly, the flesh flies, and related forms.

MOSQUITOES AND FLIES

MOSQUITOES

The mosquitoes, perhaps more than any other noxious insect, impel us to ask the impertinent question, why pests were made to annoy us. It would be well enough to answer that they were given as a test of the efficiency of our science in learning how to control them, if it were not for the other creatures, the wild animals, whose existence must be at times a continual torment from the bites of insects and from the diseases transmitted by them. Such creatures must endure their tortures without hope of relief, and there is ample evidence of the suffering that insects cause them.

In earlier and more primitive days the rainwater barrel and the town watering trough took the place of the course in nature study in our present-day schools. While the lessons of the water barrel and the trough were perhaps not exact or thoroughly scientific, we at least got our learning from them at first hand. We all knew then what "wigglers" and "horsehair snakes" were; and we knew that the former turned into mosquitoes as surely as we believed that the latter came from horsehairs. Modern nature study has set us upon the road to more exact science, but the aquarium can never hold the mysteries of the old horse trough or the marvels of the rainwater barrel.

The supposed ancestry of the horsehair snake is now an exploded myth, but the advance of science has unfortunately not altered the fact that wigglers turn into mosquitoes, except in so far as the spread of applied sanitation has brought it about that fewer of them than formerly succeed in doing so. And now, as we leave the homely objects of our first acquaintance with "wigglers" for the more convenient apparatus of the laboratory, we will call the creatures *mosquito larvae*, since that is what they are.

The rainwater barrel never told us how those wiggling

mosquito larvae got into it—that was the charm of the barrel; we could believe that we stood face to face with the great mystery of the origin of life. Now, of course, we understand that it is a very simple matter for a female

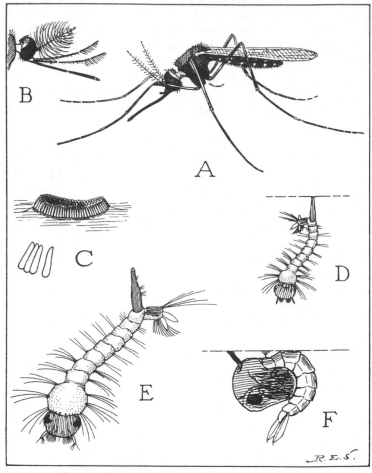

FIG. 174. Life stages of a mosquito, *Culex quinquefasciatus*
A, the adult female. B, head of an adult male. C, a floating egg raft, with four eggs shown separately and more enlarged. D, a young larva suspended at the surface of the water. E, full-grown larva. F, the pupa resting against the surface film of the water

mosquito to lay her eggs upon the surface of the water, and that the larvae come from the eggs.

There are many species of mosquitoes, but, from the standpoint of human interest, most of them are included in three groups. First there are the "ordinary" mosquitoes, species of the genus *Culex* or of related genera; second, the yellow-fever mosquito, *Aëdes aegypti*; and third, the malaria-carrying mosquitoes, which belong to the genus *Anopheles*.

The common Culex mosquitoes (Fig. 174 A) lay their eggs in small, flat masses (C) that float on the surface of the water. Each egg stands on end and is stuck close to its neighbors in such a manner that the entire egg mass has the form of a miniature raft. Sometimes the eggs toward the margin of the raft stand a little higher, giving the mass a hollowed surface that perhaps decreases the chance of accidental submergence, though the raft is buoyed up from below by a film of air beneath the eggs.

Almost any body of quiet water is acceptable to the Culex mosquito as a receptacle for her eggs, whether it be a natural pond, a pool of rainwater, or water standing in a barrel, a bucket, or a neglected tin can. Each egg raft contains two or three hundred eggs and sometimes more, but the largest raft seldom exceeds a fourth of an inch in longest diameter. The eggs hatch in a very short time, usually in less than twenty-four hours, though the incubation period may be prolonged in cool weather. The young mosquito larvae come out of the lower ends of the eggs, and at once begin an active life in the water.

The body of the young mosquito larva is slender and the head proportionately large (Fig. 174 D). As the creature becomes older, however, the thoracic region of the body swells out until it becomes as large as the head, or finally a little larger (E). The head bears a pair of lateral eyes (Fig. 175, *b*), a pair of short antennae (*Ant*), and, on the ventral surface in front of the mouth, a pair of large brushes of hairs curved inward (*a*). From the sides of the body segments project laterally groups of long hairs, some of which are branched in certain species. The rear end of the body appears to be forked, being divided into an

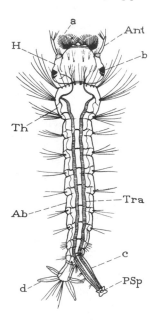

upper and a lower branch. The upper branch (*c*), however, is really a long tube projecting dorsally and backward from the next to the last segment. The lower branch is the true terminal segment of the body and bears the anal opening of the alimentary canal at its extremity. On the end of this segment four long, transparent flaps project laterally (*d*), two groups of long hairs are situated dorsally, and a fan of hairs ventrally (Fig. 174 E).

The principal characteristic of the mosquito larva is the specialization of its respiratory system. The larva breathes through a single large aperture situated on the end of the dorsal tube that projects from the next to the last segment of the body (Fig. 175, *PSp*). This orifice opens by two inner spiracles into two wide tracheal trunks (*Tra*) that extend forward in the body and give off branches to all the internal organs.

FIG. 175. Structure of a *Culex* mosquito larva

a, mouth brushes; *Ab*, abdomen; *Ant*, antenna; *b*, eye; *c*, respiratory tube; *d*, terminal lobes; *H*, head; *PSp*, posterior spiracle; *Th*, thorax; *Tra*, dorsa tracheal trunks

The mosquito larva, therefore, can breathe only when the tip of its respiratory tube projects above the surface of the water, and, though an aquatic creature, it can be drowned by long submergence. Yet the provision for breathing at the surface has a distinct advantage: it renders the mosquito larva independent of the aeration of the water it inhabits, and allows a large number of larvae to thrive in a small quantity of water, provided the latter contains sufficient food material.

The tip of the respiratory tube is furnished with five small lobes arranged like the points of a star about the central breathing hole. When the larva is below the surface, the points close over the aperture and prevent the ingress of water into the tracheae; but as soon as the tip of the tube comes above the surface, its points spread apart. Not only is the breathing aperture thus exposed, but the larva is enabled to remain indefinitely suspended from the surface film (Figs. 174 D, 181 B). In this position, with its head hanging downward, it feeds from a current of water swept toward its mouth by the vibration of the mouth brushes. Particles suspended in the water are caught on the brushes and then taken into the mouth. Any kind of organic matter among these particles constitutes the food of the larva. Larvae of Culex mosquitoes, however, feed also at the bottom of the water, where food material may be more abundant.

The body of the mosquito larva has apparently about the same density as water; when inactive below the surface, some larvae slowly sink, and others rise. But the mosquito larva is an energetic swimmer and can project itself in any direction through the water by snapping the rear half of its body from side to side, which characteristic performance has given it the popular name of "wiggler." The larva can also propel itself through the water with considerable speed without any motion of the body. This movement is produced by the action of the mouth brushes. Likewise, while hanging at the top of the water, the larva can in the same manner swing itself about on its point of suspension, or glide rapidly across the surface.

The larvae of Culex mosquitoes reach maturity in about a week after hatching, during the middle of summer; but the larval period is prolonged during the cooler seasons of spring and fall. The larva passes through three stages, and then becomes a pupa.

[333]

The mosquito pupa (Fig. 174 F) also lives in the water, but is quite a different looking creature from the larva. The thorax, the head, the head appendages, the legs, and the wings are all compressed into a large oval mass from

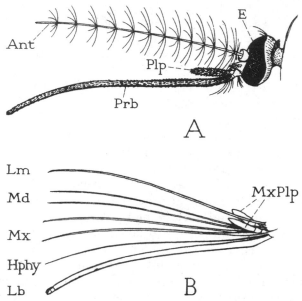

Fig. 176. Mouth parts of a female mosquito, *Joblotia digitata*

A, the head with the proboscis (*Prb*) in natural position. B, the mouth parts separated, showing the component pieces of the proboscis

Ant, antenna; *E*, compound eye; *Hphy*, hypopharynx; *Lb*, labium; *Lm*, labrum; *Md*, mandibles; *Mx*, maxillae; *MxPlp*, *Plp*, maxillary palpi; *Prb*, proboscis

which the slender abdomen hangs downward. The pupa, owing to air sacs in the thorax, is lighter than water and, when quiet, it rises to the surface where it floats with the back of the thorax against the surface film. The pupa has lost the respiratory tube and the posterior spiracles of the larva, but has acquired two large, trumpetlike breathing tubes of its own that arise from the anterior part of the thorax, the mouths of which open above the water when the pupa comes in contact with the surface. The pupa, of course, does not feed, but it is almost as active as the larva, for it must avoid its enemies. When disturbed it rapidly swims downward by quick movements of the abdomen, the extremity of which is provided with two large swimming flaps. The duration of the pupal stage in midsummer is about two days.

The adult mosquito issues from the pupal skin through a split in the back of the latter. We now see why the pupa is made lighter than water—it must float at the surface in order to allow the adult to escape into the air.

The full-fledged mosquito (Fig. 174 A) has the general features of any other two-winged fly, but it is distinguished from nearly all other flies by the presence of scales on its wings and on parts of its head, body, and appendages. The mouth parts of the adult mosquito are of the

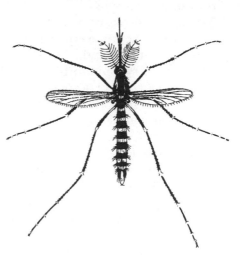

FIG. 177. *Aëdes atropalpus*, male, a mosquito related to the yellow fever mosquito and similar to it in appearance

piercing and sucking type, and are similar in structure to those of the horsefly, except that the individual pieces are longer and slenderer, and together constitute a beak, or proboscis, extending forward and downward from the head (Fig. 176 A, *Prb*). The male and the female mosquitoes are readily distinguishable by the character of the antennae, these organs in the male being large and feathery (Fig. 174 B), while those of the female are threadlike and provided with comparatively few short hairs (A). The sexes differ also in the mouth parts, for, as in the horseflies, the males lack mandibles.

The mouth parts of the mosquito, in the natural position, do not appear as separate pieces, as do those of the horsefly. The various elements, except the palpi, are compressed into a beak that projects forward and downward from the lower part of the head (Fig. 176 A, *Prb*). The length of the beak varies in different kinds of mosquitoes; it is particularly long in the large South American species shown in Figure 176.

When the beak of the female mosquito is dissected (Fig. 176 B), the same equipment of parts is revealed as is possessed by the female horsefly (Fig. 169 B), namely, a labrum (*Lm*), two mandibles (*Md*), two maxillae (*Mx*), a hypopharynx (*Hphy*), and a labium (*Lb*). It is the labium that forms most of the visible part of the beak, the other pieces being concealed within a deep groove in its upper surface.

The *labrum* (Fig. 176 B, *Lm*) is a long median blade, concave below, terminating in a hard, sharp point; it is probably the principal piercing tool of the mosquito's outfit. The *mandibles* of the mosquito (*Md*) are very slender, delicate bristles; those of the species figured are so weak that it would seem they can be of little use to the insect. The *maxillae* (*Mx*) are thin, flat organs with thickened bases, each terminating in a sharp point armed on its outer edge with a row of backward-pointing, saw-like teeth which probably serve to keep the mouth parts fixed in the puncture as the piercing labrum is thrust deeper into the flesh. The *palpi* (*MxPlp*) arise from the bases of the maxillae. The *hypopharynx* (*Hphy*) is a slender blade with a median rib which is traversed by the channel of the salivary duct. Its upper surface is concave and, in the natural position, is closed against the concave lower side of the labrum, the two apposed pieces thus forming between them a tube which leads up to the mouth opening. The saliva of the mosquito is injected into the wound from the tip of the hypopharynx, and the blood of the victim is sucked up to the mouth through the labro-hypopharyngeal tube. The labium (*Lb*) serves

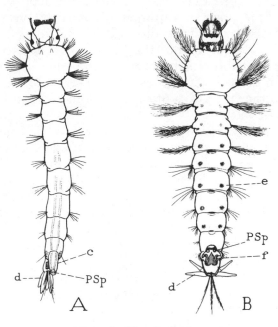

FIG. 178.   Mosquito larvae

A, *Aëdes atropalpus*.   B, *Anopheles punctipennis*, the malaria mosquito larva

c, respiratory tube; *d*, terminal lobes; *e*, stellate groups of hairs that hold the larva at the surface of the water (fig. 181 A); *f*, spiracular area; *PSp*, spiracle

principally as a sheath for the other organs. It ends in two small lateral lobes, the labella, between which projects a weak, median tonguelike process. When the mosquito pierces its victim the base of the labium bends backward as the other bristlelike members of the group of mouth parts sink into the wound.

Mosquitoes of both sexes are said to feed on the sap of

plants, which they extract by puncturing the plant tissues; they will also feed on the exuding juices of fruit, or on any soft vegetable matter. The females, however, are notorious for their propensity for animal blood, and they by no means limit their quest for this article of food to human beings. The male mosquitoes, apparently, very rarely depart from a vegetarian diet. The pain from the bite of a female mosquito and the subsequent irritation and swelling probably result from the injection of the secretion from the salivary glands of the insect into the wound. It is said that the saliva of the mosquito prevents coagulation of the blood.

Because of the short time necessary for the completion of the life cycle from egg to adult during summer, there are many generations of mosquitoes from spring to fall. The winter is passed both in the adult and in the larval stage. Fertile females may survive cold weather in protected places; and larvae found in large numbers, frozen solid in the ice of ponds, have become active on being thawed out, and capable of development when given a sufficient degree of warmth.

The yellow-fever mosquito, now known as *Aëdes aegypti* but at the time of the discovery of its relation to yellow fever generally called *Stegomyia fasciata*, is similar in its habits during the larval and pupal stages to the Culex mosquitoes. It lays its eggs singly, however, and they float unattached on the surface of the water. The adult mosquito may be identified by its decorative markings. On the back of the thorax is a lyrelike design in white on a black ground; the joints of the legs are ringed with white; the black abdomen is conspicuously cross-banded with white on the basal half of each segment. The male has large plumose antennae and long maxillary palpi. The female has a strong beak, but small palpi, and her antennae are of the short-haired form usual with female mosquitoes. The species of *Aëdes* shown in Figure 177 much resembles the yellow-fever mosquito, but it is a more northern one common about Washington, D. C., where it breeds in rock pools along the Potomac River.

The larva of Aëdes (Fig. 178 A) resembles a Culex larva, but it feeds more habitually at the bottom of the water and may spend long periods below without coming to the

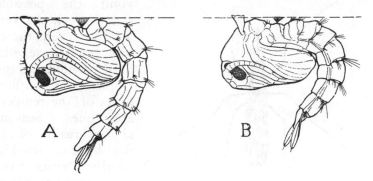

FIG. 179.   Mosquito pupae in natural position resting against the under surface of the water

A, *Aëdes atropalpus*. B, *Anopheles punctipennis*

surface for air.   In its search for food it noses about in the refuse at the bottom of the water and voraciously consumes dead insects and small crustaceans.   The pupa likewise (Fig. 179 A) does not differ materially from a Culex pupa.   When quiet it floats at the surface of the water with the entire back of its thorax against the surface film and the tips of its breathing tubes above the surface. Probably no mosquito pupa hangs suspended from its respiratory tubes in the manner in which the pupae of various species are often figured.

Aëdes aegypti is the only known natural carrier of the virus of yellow fever from one person to another.   The disease can be taken only from the bite of a mosquito of this species that has become infected by previous feeding on the blood of a yellow-fever patient.   The organism that produces yellow fever is perhaps not yet definitely known, though strong evidence has been adduced to show that it is one of the minute, non-filterable organisms called *spirochetes*. The virus will not develop in the mosquitoes at a temperature below 68° F., and *Aëdes aegypti* will not breed in latitudes much beyond the possible range of yellow fever. Yellow fever, therefore, is a disease ordinarily confined to the tropics and warmer parts of the temperate zones. Seasonal outbreaks of it that have occurred in northern cities have been caused probably by local infestations of infected mosquitoes brought in on ships from some southern port.

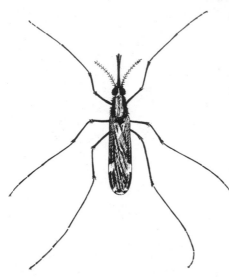

FIG. 180. The female malaria mosquito, *Anopheles punctipennis*

The malaria mosquitoes belong to the genus *Anopheles*, a genus represented by species in most temperate and tropical regions of the world, which are prevalent wherever malaria occurs. Our most common malaria species is *Anopheles punctipennis* (Fig. 180), characterized by a pair of dull white spots on the edges of the wings. The *Anopheles* females lay their eggs singly on the surface of the water, where they float, each buoyed up by an air jacket about its middle.

The larvae of Anopheles (Fig. 178 B) differ conspicuously from those of Culex and Aëdes both in structure and habits. Instead of a respiratory tube projecting from near the end of the body, as in Culex (Figs. 174 E, 175), there is a concave disc (Fig. 178 B, *f*) on the back of the next to the last segment, in which the posterior spiracles (*PSp*) are located. The larva floats in a horizontal position just below the surface film of the water (Fig. 181 A), from which it is suspended by a series of floats (Fig. 178 B, *e*) consisting of starlike groups of short hairs arranged in pairs along the back. The spreading tips of the hairs pro-

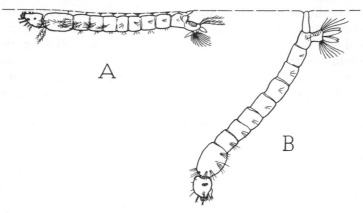

FIG. 181. Feeding positions of *Anopheles* and *Culex* mosquito larvae
A, *Anopheles* larva suspended horizontally beneath the surface film, and feeding at the surface with its head inverted. B, *Culex* larva hanging from the respiratory tube

ject slightly above the water surface and keep the larva afloat. In the floating position, the respiratory disc breaks through the surface film, and its raised edges leave a dry area surrounding the spiracles. The long hairs that project from the sides of the thorax and the first three body segments are mostly branched and plumose.

The Anopheles larva (Fig. 181 A) feeds habitually at the top of the water. When disturbed it shoots rapidly across the surface in any direction, but goes downward reluctantly. In order to feed in its horizontal position, it turns its head completely upside down and with its mouth brushes creates a surface current toward its mouth.

The pupa of Anopheles (Fig. 179 B) is not essentially

different from that of Culex or Aëdes. Its most distinctive character is in the shape of the respiratory tubes, which are very broad at the ends.

The parasite of malaria is not a bacterium but a microscopic protozoan animal named *Plasmodium*. There are several species or varieties that correspond with the different varieties of the disease. The malaria Plasmodium has a complicated life cycle and is able to complete its life only when it can spend a part of it in the body of a mosquito and the other part in some vertebrate animal. In the human body the malaria parasites live in the red corpuscles of the blood. Here they multiply by asexual reproduction, producing for a while many other asexual generations. Eventually, however, certain individuals are formed that, if taken into the stomach of an Anopheles mosquito, develop there into males and females. In the stomach of the mosquito, these sexual individuals unite in pairs, and the resulting *zygotes*, as they are called, penetrate into the cells of the stomach wall. Here they live for a while and multiply into a great number of small spindle-shaped creatures, which go through the stomach wall into the body cavity of the mosquito and at last collect in the salivary glands. If now the mosquito, with its salivary glands full of the Plasmodium parasites in this stage, bites some other animal, the parasites are almost sure to be injected into the wound with the saliva. If they are not at once destroyed by the white blood corpuscles, they will quickly enter the red blood corpuscles, and the victim will soon show symptoms of malaria.

The House Fly and Some of Its Relations

Our familiar domestic pest, the house fly, may be taken as the type of a large group of flies, and in particular of those belonging to the family Muscidae, which is named from its best known member, *Musca domestica*, the house fly—*musca* being the Latin word for fly.

The house fly (Fig. 182 A), though particularly a domestic pest to people that live indoors, is intimately associated with the stable. Its favorite breeding place is the manure pile. Here the female fly lays her eggs (B), and here the larvae, or maggots (C), live until they are ready for transformation. It is estimated that fully ninety-five per cent of our house flies have been bred in horse manure. A few may come from garbage cans, or from heaps of vegetable refuse, but such sources of fly infestation are comparatively unimportant. Measures of fly control are directed chiefly to preventing the access of flies to stable manure and the destruction of maggots living in it.

The eggs of the house fly (Fig. 182 B) are small, white, elongate-oval objects, about one twenty-fifth of an inch in length, each slightly curved on one side and concave on the other. The female fly begins to lay eggs in about ten days after having transformed to the adult form, and she deposits from 75 to 150 eggs at a single laying. She repeats the laying, however, at intervals during her short productive period of about twenty days, and in all may deposit over 2,000 eggs. Each egg hatches in twenty-four hours or less.

The larva of the house fly, in common with that of many other related flies, is a particularly wormlike creature, and is commonly called a *maggot* (Fig. 182 D). Its slender white body is segmented, but, in external appearance, it is legless and headless. On a flat area at the rear end of the body are located two large spiracles (*PSp*), which the novice might mistake for eyes. The tapering end of the body is the head end, but the true head of the maggot is withdrawn entirely into the body. From the aperture where the head has disappeared, which serves the maggot as a mouth, two clawlike hooks project (*mh*), and these hooks are both jaws and grasping organs to the maggot. The larva sheds its skin twice during the active part of its life, which is very short, usually only two or three weeks. Then it crawls off to a secluded place, generally in the earth beneath its manure pile, where it enters a resting condition. Its skin now hardens and contracts until the creature takes on the form of a small, hard-shelled, oval capsule, called a *puparium* (Fig. 182 E).

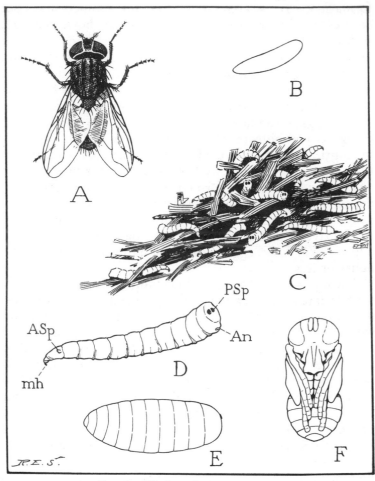

Fig. 182. The house fly, *Musca domestica*

A, the adult fly (5½ times natural size). B, the house fly egg (greatly magnified). C, larvae, or maggots, in manure. D, a larva (more enlarged). E, the puparium, or hardened larval skin which becomes a case in which the larva changes to a pupa. F, the pupa

Within the puparium, the larva sheds another skin, and then transforms to the pupa. The pupa (Fig. 182 F) is thus protected during its transformation to the adult by the puparial skin of the larva, which serves in place of a cocoon. When the adult is fully formed, it pushes off a circular cap from the anterior end of its case, and the fly emerges. The length of the entire cycle from egg to adult varies according to temperature conditions, but it is usually from twelve to fourteen days. The adult flies are short-lived in summer, thirty days, or not more than two months, being their usual span of life. In cooler weather, however, when their activities are suppressed, they live longer, and a few survive the winter in protected places.

One of the essential differences between flies of the house fly type and the mosquitoes and horseflies is in the structure of the mouth parts. The house fly lacks mandibles and maxillae, but it retains the median members of the normal group of mouth-part pieces, which are the labrum, the hypopharynx, and the labium. These parts are combined to form a sucking proboscis that is ordinarily folded beneath the head, but which is extended downward when in use (Fig. 183 A, *Prb*).

The labium (Fig. 183 B, *Lb*) is the principal component of the proboscis of the house fly, and its terminal lobes, or labella (*La*), are particularly well developed. From the base of the labium there projects forward a pair of palps (*Plp*), which are probably the palpi of the maxillae, though those organs are otherwise lacking. The anterior surface of the labium is deeply concave, but its trough-like hollow is closed by the labrum (*Lm*). Against the labial wall of the inclosed channel lies the hypopharynx (*Hphy*). When the lobes of the labium are spread out, the anterior cleft between them is closed except for a small central aperture (*a*). This opening becomes the functional mouth of the fly, though the true mouth is situated, as in other insects, between the bases of the labrum and the hypopharynx, and opens into a large sucking pump

having the same essential structure as that of the horsefly (Fig. 170 A).

The house fly has no piercing organs; it subsists entirely on a liquid diet. The food liquid enters the aperture between the labella, and is drawn up to the true

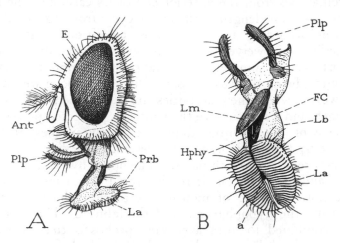

Fig. 183.   Head and mouth parts of the house fly

A, lateral view of the head with the proboscis (*Prb*) extended. *Ant*,
antenna; *E*, compound eye; *La*, labella, terminal lobes of the proboscis; *Plp*, maxillary palpi (the maxillae are lacking); *Prb*, proboscis

B, the proboscis of the fly, as seen in three-quarter front view and
from below.   The proboscis consists of the thick labium (*Lb*), ending
in the labellar lobes (*La*), between which is a small pore (*a*) leading
into the food canal (*FC*) of the proboscis.   The food canal contains
the hypopharynx (*Hphy*), and is closed in front by the labrum (*Lm*)

mouth through the food canal in the labium between the
labrum and the hypopharynx. The fly, however, is not
dependent on natural liquids; it can dissolve soluble substances, such as sugar, by means of its saliva. The
saliva is ejected from the tip of the hypopharynx, and
probably spreads over the food through the channels of
the labial lobes. These same channels, perhaps, also
collect the food solution and convey it to the labellar
aperture.

MOSQUITOES AND FLIES

During recent years we have become so well educated concerning the ways of the house fly, its disgusting habits of promiscuous feeding, now in the garbage can or somewhere worse, and next at our table or on the baby's face, and we have learned so much about its menace as a possible carrier of disease, that it is scarcely necessary to enlarge here upon the fly's undesirability as a domestic companion.

The most serious accusation against the house fly is that, owing to the many kinds of places it frequents without regard to sanitary conditions, and to its indiscriminate feeding habits, there is always a chance of its feet, body, mouth parts, and alimentary canal being contaminated with the germs of disease, particularly those of typhoid fever, tuberculosis, and dysentery. It has been demonstrated that flies can carry germs about with them which will grow when given a proper medium, and likewise that flies taken at large may be covered with bacteria, a single fly sometimes being loaded with millions of them. The wisdom of sanitary measures for the protection of food from contamination by flies can not, therefore, be questioned.

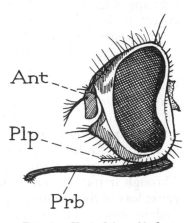

FIG. 184. Head of the stable fly, *Stomoxys calcitrans*

Ant, antenna; *Plp*, maxillary palpus; *Prb*, proboscis

There is one form of insect villainy, however, of which the house fly is not guilty; the structure of its mouth parts clears it of all accusations of biting. And yet we hear it often asserted by persons of unquestioned veracity that they have been bitten by house flies. The case is one of mistaken identification and not of imagination on the part of the plaintiff; the insect that inflicts the bite is not the house fly, but another species closely resembling the common domestic fly in general appearance, though a little smaller. If the culprit is caught, there may be seen projecting from its head a long, hard, tapering beak (Fig. 184, *Prb*), an organ quite different from any part of the mouth equipment of the true house fly (Fig. 183). This biting fly is, in fact, the *stable fly*, a species known to entomologists as *Stomoxys calcitrans*. It belongs to the same family as the house fly, and while it sometimes comes about houses, it is particularly a pest of horses and cattle.

The stable fly lives in most parts of the inhabited world. Both sexes have blood-sucking habits, and probably feed on any kind of warm-blooded animal, though the species is most familiar as a frequenter of stables and as a pest of domestic stock. The stable fly breeds mostly in fermenting vegetable matter, the larvae being found principally under piles of wet straw, hay, alfalfa, grain, weeds, or any vegetable refuse.

Cattle are afflicted by another pestiferous fly called the horn fly, or *Haematobia irritans*. The species gets its common name from the fact that it is usually observed about the bases of the horns of cattle, where great numbers of individuals often assemble. But the horns of the animals are merely convenient resting places. Haematobia is a biting fly like Stomoxys, and, because of its greater numbers, it often becomes a most serious pest of cattle. Through irritation and annoyance during feeding, it may cause loss of flesh in grazing stock, and a reduction of milk in dairy cows. The horn fly resembles the stable fly, but is smaller, being about one-half the size of the house fly. It breeds mostly in fresh manure of cattle dropped in the fields.

Of all the biting flies there is none to compare with the tsetse fly of Africa (Fig. 185). Not only is this fly an intolerable nuisance to men and animals because of the severity of its bite, but it is a deadly menace by reason of its being the carrier of the parasite of African sleeping sickness of man, and that of the related disease called *nagana* in horses and cattle.

African sleeping sickness is caused by a protozoan parasite of the genus *Trypanosoma* that lives in the blood and other body liquids. Trypanosomes are active, one-celled organisms having one end of the body prolonged into a tail, or flagellum. They are found as parasites in many vertebrate animals, but most of them do not produce disease conditions. There are at least three African species, however, whose presence in the blood of their hosts means almost certain death. Two cause the sleeping sickness in man, and the other produces nagana in horses, mules, and cattle. The two human species have different distributions and produce each a distinct variety of the disease. One is confined to the tropical parts of Africa, the other is more southern. The southern form of the disease is said to be much more severe than the tropical form, claiming its victims in a matter of months, while the other may drag along for years. The sleeping sickness and nagana trypanosomes are entirely dependent in nature on the tsetse flies for their means of transport from one person or from one animal to another.

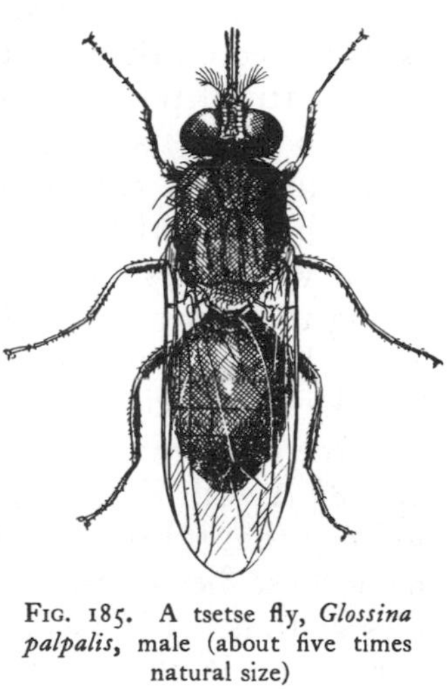

FIG. 185. A tsetse fly, *Glossina palpalis*, male (about five times natural size)

The tsetse fly (Fig. 185) is a larger relation of the horn fly and the stable fly, having the same type of beak and an insatiable appetite for blood. The tsetse fly genus is *Glossina*. There are two species particularly concerned with the transportation of sleeping sickness, corresponding with the two species of trypanosomes that cause the two forms of the disease. One is *Glossina palpalis* (Fig. 185), distributor of the tropical variety of the disease; the other is *Glossina morsitans*, carrier both of the southern variety of sleeping sickness and of nagana.

The stable fly, the horn fly, and the tsetse fly, we have said, belong to the same family as the house fly, namely, the Muscidae; and yet they appear to have mouth parts of a very different type. The differences, however, are of a superficial nature. All the muscid flies, biting and non-biting, have the same mouth-part pieces, which are the labrum (Figs. 183 B, 186 C, *Lm*), the hypopharynx (*Hphy*), and the labium (*Lb*). They lack mandibles and maxillae, though the maxillary palps (*Plp*) are retained. In the biting species, the labium is drawn out into a long, slender rod (Fig. 186 C, *Lb*), and its terminal lobes, the labella (*La*), are reduced to a pair of small, sharp-edged plates armed on their inner surfaces with teeth and ridges. In the natural position, the deflected edges of the labrum (Fig. 186 B, *Lm*) are held securely within the hollow of the upper surface of the labium (*Lb*), the two parts thus inclosing between them a large food canal (*FC*) at the bottom of which lies the slender hypopharynx (*Hphy*), containing the exit tube of the salivary duct.

The biting muscids, therefore, have a strong, rigid, beaklike proboscis formed of the same pieces that compose the sucking proboscis of the house fly (compare Fig. 183 A with Figs. 184 and 186 A), but the labium is so modified that it becomes an effective piercing organ. When one of these flies bites, it sinks the entire beak into the flesh of its victims. The tsetse fly is said to spread its front legs apart when it alights for the purpose of feeding, and to insert its beak by several quick downward thrusts of the head and thorax. The insect then quickly fills itself with blood, with which it may become so distended that it can scarcely fly. The bulb at the base of the tsetse fly's labium (Fig. 186 C, *b*) is no part of the sucking apparatus; it is merely an enlargement for the accommodation of muscles. The true sucking organ lies within the head (*Pmp*), and does not differ in structure from that of other flies.

While our indictment of the flies has applied thus far only to the insects in the mature form, there are species which, though entirely innocent of any criminality in their

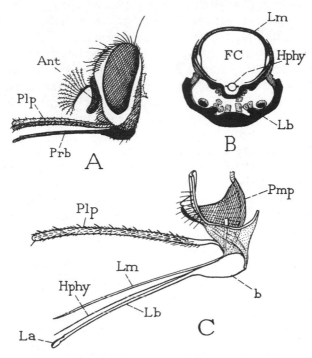

Fig. 186. Head and mouth parts of the tsetse fly, *Glossina*

A, lateral view of the head and proboscis (*Prb*) of *Glossina palpalis*, male

B, cross-section of the proboscis of *Glossina fusca* (from Vogel), showing the food canal (*FC*) inclosed by the labrum (*Lm*) and labium (*Lb*), and containing the tubular hypopharynx (*Hphy*) through which the saliva is injected into the wound

C, mouth parts of *Glossina palpalis*, with the parts of the proboscis separated. *b*, basal swelling of the labium; *La*, the labella, or terminal lobes of the labium used for cutting into the skin of the victim; *Lb*, labium; *Lm*, labrum; *Plp*, maxillary palpus (the maxillae are lacking); *Pmp*, mouth pump

adult behavior, are, however, most obnoxious creatures during their larval stages. The ordinary blowflies, which are related to the house fly, lay their eggs in the bodies of dead animals, where the larvae speedily hatch and feed on the putrefying flesh. Another kind of blowfly deposits living larvae instead of eggs. These flies may be regarded as beneficial in that their larvae are scavengers. But some of their relations appear to have taken a diabolical hint from their habits, for they make a practice of depositing their eggs in open wounds, sores, or in the nostrils of living animals, including man. The larvae burrow into the tissues of the victims and cause extreme annoyance, suffering, and even death. A notable species of this class of pests is the screw worm. Infestation by fly larvae, or maggots, is called *myiasis*.

Well-known cases of animal myiasis are that of the botfly in horses and of the ox warble in cattle. The flies of both these species lay their eggs on the outside of the animals. The young larvae of the botfly are licked off and swallowed, and then live until full-grown in the stomach of the host. The young ox-warble larva burrows into the flesh of its host and lives in the body tissues until mature, when it bores through the skin on the back of the afflicted beast, drops out, and completes its transformation in the ground.

Not only animals but plants as well are subject to internal parasitism by fly larvae. Garden crops are attacked by leaf maggots and root maggots; orchardists in the northern States have to contend against the apple maggot, which is a relation of the olive fly of southern Europe and of the destructive fruit flies of tropical countries. That notorious scourge of wheat fields, the Hessian fly, is a second or third cousin of the mosquito, and it is in its larva form that it makes all the trouble.

The special attention that has been given to pestiferous flies must make it appear that the Diptera are a most undesirable order of insects. As a matter of fact, however, there are thousands of species of flies that do not affect us in any injurious way; while, furthermore, there are species, and many of them, that render us a positive service by the fact that their larvae live as parasites in the bodies of other injurious insects and bring about the destruction of large numbers of the latter.

Scientifically, the Diptera are most interesting insects, because they illustrate more abundantly than do the members of any other order the steps by which nature has achieved evolution in animal forms. An entomologist would say that the Diptera are highly specialized insects; and as evidence of this statement he would point out that the flies have developed the mechanical possibilities of the common insect mechanism to the highest general level of efficiency attained by any insect and that they have carried out many lines of special modification, giving a great variety of new uses for structures originally limited to one mode of action. But when we say that any animal has developed to this or that point of perfection, we do not mean just what we say, for the creature itself has been the passive subject of influences working upon it or within it. A fundamental study of biology in the future will consist of an attempt to discover the forces that bring about evolution in living things.

INDEX

INDEX

INDEX

[357]

INDEX

House fly, pupa, 345
  puparium, 344
  unsanitary habits, 347
Hypermetamorphosis, 250
Hyperparasite, 181
Hypopharynx, 108

I

Imaginal discs, 259
Imago, defined, 259
Insecta, 28
Intestine, 110

J

Jumping bush cricket, 69
  song, 70
June bugs, 230

K

Katydid, 43
  habits, 46
  musical instruments, 47, 48
  song, 44, 48, 49
  true, 43
Katydid family, 32
Katydids, 32
  angular-winged, 41
  bush, 38
  ears, 36
  musical instruments, 34–36
  round-headed, 37
  song, 33, 34
  young, 11
King termite, 134

L

Labium, 108
Labrum, 108
Lady-beetles, 175, 230
Ladybird beetles, 175, 230
Larva, characters, 246
  definition, 245
  nature, 249
  of *Aëdes*, 339
    Anopheles, 340, 341

Larva of *Culex*, 331, 332, 333
  of flies, 325
    house fly, 343
    mosquitoes, 329
    wasps and bees, 252
Leaf insect, 71, 72, 73
Legs of insects, 107
Lepisma, 93
Life of a caterpillar, 262
Locustidae, 32
Locusts, 2
  seventeen-year, 182
Luna moth, 228, 230

M

Machilis, 93
Maggots, 252
Malacosoma americana, see Tent
  caterpillar
Malaria mosquitoes, 340
  adult, 340
  eggs, 340
  larva, 340, 341
  pupa, 341
Malaria parasite, 342
Malpighian tubules, 116
Mandibles, 107
Mantids, 73
Mantis, praying, 73–76
  eggs, 75, 76
Maxillae, 108
May beetles, 230
Mayfly, 96
Meadow grasshoppers, 52–54
Mecostethus gracilis, 31
Metabolism of pupa, 260
Metamorphosis, 14, 226
  complete, 245
  defined, 227
  diagram, 243
  incomplete, 245
  of tent caterpillar, 297, 299–304
Microcentrum retinerve, 41, 43
  song, 43

INDEX

INDEX

Periodical cicada, air chamber, 205
 broods, 215–217
 death of adults, 214
 digging methods, 190, 191
 egg laying, 212–214
 eggs, 212, 219
 food, 200
 front leg of nymph, 190
 hatching of eggs, 217–223
 head of adult, 201
 huts, turrets, 192
 mouth parts, 201–205
 musical instruments, 199, 207–212
 nymphal chambers, 187–189
  stages, 186, 187
 nymphs, 185–193, 223–225
 ovipositor, 199
 races, 215
 salivary pump, 204
 song of large variety, 210, 211
  of small variety, 211, 212
 sucking mechanism, 203
 transformation, 193–199
 two varieties, 199
 young nymphs, 223–225
Phagocytes, 259, 301
Phaneroptera, 38
Pharynx, 110
Phylloxera, 172
Phylum, 26
Physiology of tent caterpillar, 283
Plant lice, 152
Plasmodium, 342
Proboscis of moth, 307, 308
Promethea moth, 228, 229
Propupa of tent caterpillar, 296–298
Protoplasm, 100
Pterophylla camellifolia, see Katydid
Pupa, 250, 253, 254
 added stage in metamorphosis, 254

Pupa, definition, 245
 of flies, 327
  house fly, 345
  mosquitoes, 334, 339, 341
  tent caterpillar, 298
 reason for, 257
Puparium, 252
 of house fly, 344

Q

Queen termite, 134, 149

R

Rat-tailed maggot, 327
Reproduction, 102
Reproductive organs, 122
Respiration, 114
Reticulitermes, 136
 life history, 136–141
Rhadophorinae, 55
Roaches, 77, 80
 and other ancient insects, 77
 eggs, 80, 81
Robber flies, 324
Rocky Mountain locust, 17, 18, 19
Rosy apple aphis, 168–170
Round-headed katydids, 37
 Amblycorypha oblongifolia, 39
 angular-winged, 41
 fork-tailed bush, 39
 Microcentrum, 41, 43
 Phaneroptera, 38
 Scudderia, 38, 39

S

Sarcophaga kellvi, 19–21
Scudderia, 38
 furcata, see Fork-tailed bush katydid
Segments of body, 12
Sense organs, 121

[360]

INDEX

INDEX

Termites, wingless males and
  females, 140
  workers, 131
  young, 136
Testes, 122
Tettigoniidae, 32
Thorax, 12
Thysanura, 247
Tracheae, 114
Tree crickets, 63
  antennal marks, 66, 67
  attraction of males for females,
    68, 69
  black-horned, 67
  broad-winged, 65
  four-spotted, 67
  musical organs, 56, 57
  narrow-winged, 67
  Neoxabia, 69
  Oecanthus, 65–68
  snowy, 65
  song, 65, 66, 67
  two-spotted, 69
Triungulins, 23
Tropisms, 121
True katydid, 43
Trypanosoma, 349
Tsetse fly, 348, 349, 350
  mouth parts, 350, 351
Two-spotted tree cricket, 69
  song, 69

W

Walking stick, 72
Walking stick insects, 71
Wasps and bees, larvae, 230, 238,
  252
Ways and means of living, 99
White ants, 128
White grubs, 230
Wings, 83, 84
  evolution, 315
  of bees, 319
    beetles, 318
    butterflies and moths, 318
    dragonflies, 316
    flies, 319
    grasshoppers, 318
    roaches, 83, 84, 318
    termites, 146, 316
    wasps, 319
  origin, 91, 92
Wigglers, 230, 329
Woolly aphis, 172

X

Xiphidium, 54

Y

Yellow fever, 339
Yellow fever mosquito, 331, 339,
  340

A CATALOGUE OF SELECTED DOVER BOOKS
IN ALL FIELDS OF INTEREST

A CATALOGUE OF SELECTED DOVER BOOKS
IN ALL FIELDS OF INTEREST

AMERICA'S OLD MASTERS, James T. Flexner. Four men emerged unexpectedly from provincial 18th century America to leadership in European art: Benjamin West, J. S. Copley, C. R. Peale, Gilbert Stuart. Brilliant coverage of lives and contributions. Revised, 1967 edition. 69 plates. 365pp. of text.

21806-6 Paperbound $3.00

FIRST FLOWERS OF OUR WILDERNESS: AMERICAN PAINTING, THE COLONIAL PERIOD, James T. Flexner. Painters, and regional painting traditions from earliest Colonial times up to the emergence of Copley, West and Peale Sr., Foster, Gustavus Hesselius, Feke, John Smibert and many anonymous painters in the primitive manner. Engaging presentation, with 162 illustrations. xxii + 368pp.

22180-6 Paperbound $3.50

THE LIGHT OF DISTANT SKIES: AMERICAN PAINTING, 1760-1835, James T. Flexner. The great generation of early American painters goes to Europe to learn and to teach: West, Copley, Gilbert Stuart and others. Allston, Trumbull, Morse; also contemporary American painters—primitives, derivatives, academics—who remained in America. 102 illustrations. xiii + 306pp. 22179-2 Paperbound $3.00

A HISTORY OF THE RISE AND PROGRESS OF THE ARTS OF DESIGN IN THE UNITED STATES, William Dunlap. Much the richest mine of information on early American painters, sculptors, architects, engravers, miniaturists, etc. The only source of information for scores of artists, the major primary source for many others. Unabridged reprint of rare original 1834 edition, with new introduction by James T. Flexner, and 394 new illustrations. Edited by Rita Weiss. $6\frac{5}{8}$ x $9\frac{5}{8}$.

21695-0, 21696-9, 21697-7 Three volumes, Paperbound $13.50

EPOCHS OF CHINESE AND JAPANESE ART, Ernest F. Fenollosa. From primitive Chinese art to the 20th century, thorough history, explanation of every important art period and form, including Japanese woodcuts; main stress on China and Japan, but Tibet, Korea also included. Still unexcelled for its detailed, rich coverage of cultural background, aesthetic elements, diffusion studies, particularly of the historical period. 2nd, 1913 edition. 242 illustrations. lii + 439pp. of text.

20364-6, 20365-4 Two volumes, Paperbound $6.00

THE GENTLE ART OF MAKING ENEMIES, James A. M. Whistler. Greatest wit of his day deflates Oscar Wilde, Ruskin, Swinburne; strikes back at inane critics, exhibitions, art journalism; aesthetics of impressionist revolution in most striking form. Highly readable classic by great painter. Reproduction of edition designed by Whistler. Introduction by Alfred Werner. xxxvi + 334pp.

21875-9 Paperbound $2.50

How to Know the Wild Flowers, Mrs. William Starr Dana. This is the classical book of American wildflowers (of the Eastern and Central United States), used by hundreds of thousands. Covers over 500 species, arranged in extremely easy to use color and season groups. Full descriptions, much plant lore. This Dover edition is the fullest ever compiled, with tables of nomenclature changes. 174 full-page plates by M. Satterlee. xii + 418pp. 20332-8 Paperbound $2.75

Our Plant Friends and Foes, William Atherton DuPuy. History, economic importance, essential botanical information and peculiarities of 25 common forms of plant life are provided in this book in an entertaining and charming style. Covers food plants (potatoes, apples, beans, wheat, almonds, bananas, etc.), flowers (lily, tulip, etc.), trees (pine, oak, elm, etc.), weeds, poisonous mushrooms and vines, gourds, citrus fruits, cotton, the cactus family, and much more. 108 illustrations. xiv + 290pp. 22272-1 Paperbound $2.50

How to Know the Ferns, Frances T. Parsons. Classic survey of Eastern and Central ferns, arranged according to clear, simple identification key. Excellent introduction to greatly neglected nature area. 57 illustrations and 42 plates. xvi + 215pp. 20740-4 Paperbound $2.00

Manual of the Trees of North America, Charles S. Sargent. America's foremost dendrologist provides the definitive coverage of North American trees and tree-like shrubs. 717 species fully described and illustrated: exact distribution, down to township; full botanical description; economic importance; description of subspecies and races; habitat, growth data; similar material. Necessary to every serious student of tree-life. Nomenclature revised to present. Over 100 locating keys. 783 illustrations. lii + 934pp. 20277-1, 20278-X Two volumes, Paperbound $6.00

Our Northern Shrubs, Harriet L. Keeler. Fine non-technical reference work identifying more than 225 important shrubs of Eastern and Central United States and Canada. Full text covering botanical description, habitat, plant lore, is paralleled with 205 full-page photographs of flowering or fruiting plants. Nomenclature revised by Edward G. Voss. One of few works concerned with shrubs. 205 plates, 35 drawings. xxviii + 521pp. 21989-5 Paperbound $3.75

The Mushroom Handbook, Louis C. C. Krieger. Still the best popular handbook: full descriptions of 259 species, cross references to another 200. Extremely thorough text enables you to identify, know all about any mushroom you are likely to meet in eastern and central U. S. A.: habitat, luminescence, poisonous qualities, use, folklore, etc. 32 color plates show over 50 mushrooms, also 126 other illustrations. Finding keys. vii + 560pp. 21861-9 Paperbound $3.95

Handbook of Birds of Eastern North America, Frank M. Chapman. Still much the best single-volume guide to the birds of Eastern and Central United States. Very full coverage of 675 species, with descriptions, life habits, distribution, similar data. All descriptions keyed to two-page color chart. With this single volume the average birdwatcher needs no other books. 1931 revised edition. 195 illustrations. xxxvi + 581pp. 21489-3 Paperbound $4.50

AMERICAN FOOD AND GAME FISHES, David S. Jordan and Barton W. Evermann. Definitive source of information, detailed and accurate enough to enable the sportsman and nature lover to identify conclusively some 1,000 species and sub-species of North American fish, sought for food or sport. Coverage of range, physiology, habits, life history, food value. Best methods of capture, interest to the angler, advice on bait, fly-fishing, etc. 338 drawings and photographs. 1 + 574pp. 6⅝ x 9⅜.

22383-1 Paperbound $4.50

THE FROG BOOK, Mary C. Dickerson. Complete with extensive finding keys, over 300 photographs, and an introduction to the general biology of frogs and toads, this is the classic non-technical study of Northeastern and Central species. 58 species; 290 photographs and 16 color plates. xvii + 253pp.

21973-9 Paperbound $4.00

THE MOTH BOOK: A GUIDE TO THE MOTHS OF NORTH AMERICA, William J. Holland. Classical study, eagerly sought after and used for the past 60 years. Clear identification manual to more than 2,000 different moths, largest manual in existence. General information about moths, capturing, mounting, classifying, etc., followed by species by species descriptions. 263 illustrations plus 48 color plates show almost every species, full size. 1968 edition, preface, nomenclature changes by A. E. Brower. xxiv + 479pp. of text. 6½ x 9¼.

21948-8 Paperbound $5.00

THE SEA-BEACH AT EBB-TIDE, Augusta Foote Arnold. Interested amateur can identify hundreds of marine plants and animals on coasts of North America; marine algae; seaweeds; squids; hermit crabs; horse shoe crabs; shrimps; corals; sea anemones; etc. Species descriptions cover: structure; food; reproductive cycle; size; shape; color; habitat; etc. Over 600 drawings. 85 plates. xii + 490pp.

21949-6 Paperbound $3.50

COMMON BIRD SONGS, Donald J. Borror. 33⅓ 12-inch record presents songs of 60 important birds of the eastern United States. A thorough, serious record which provides several examples for each bird, showing different types of song, individual variations, etc. Inestimable identification aid for birdwatcher. 32-page booklet gives text about birds and songs, with illustration for each bird.

21829-5 Record, book, album. Monaural. $2.75

FADS AND FALLACIES IN THE NAME OF SCIENCE, Martin Gardner. Fair, witty appraisal of cranks and quacks of science: Atlantis, Lemuria, hollow earth, flat earth, Velikovsky, orgone energy, Dianetics, flying saucers, Bridey Murphy, food fads, medical fads, perpetual motion, etc. Formerly "In the Name of Science." x + 363pp.

20394-8 Paperbound $2.00

HOAXES, Curtis D. MacDougall. Exhaustive, unbelievably rich account of great hoaxes: Locke's moon hoax, Shakespearean forgeries, sea serpents, Loch Ness monster, Cardiff giant, John Wilkes Booth's mummy, Disumbrationist school of art, dozens more; also journalism, psychology of hoaxing. 54 illustrations. xi + 338pp.

20465-0 Paperbound $2.75

THE PRINCIPLES OF PSYCHOLOGY, William James. The famous long course, complete and unabridged. Stream of thought, time perception, memory, experimental methods—these are only some of the concerns of a work that was years ahead of its time and still valid, interesting, useful. 94 figures. Total of xviii + 1391pp.
20381-6, 20382-4 Two volumes, Paperbound $8.00

THE STRANGE STORY OF THE QUANTUM, Banesh Hoffmann. Non-mathematical but thorough explanation of work of Planck, Einstein, Bohr, Pauli, de Broglie, Schrödinger, Heisenberg, Dirac, Feynman, etc. No technical background needed. "Of books attempting such an account, this is the best," Henry Margenau, Yale. 40-page "Postscript 1959." xii + 285pp.
20518-5 Paperbound $2.00

THE RISE OF THE NEW PHYSICS, A. d'Abro. Most thorough explanation in print of central core of mathematical physics, both classical and modern; from Newton to Dirac and Heisenberg. Both history and exposition; philosophy of science, causality, explanations of higher mathematics, analytical mechanics, electromagnetism, thermodynamics, phase rule, special and general relativity, matrices. No higher mathematics needed to follow exposition, though treatment is elementary to intermediate in level. Recommended to serious student who wishes verbal understanding. 97 illustrations. xvii + 982pp.
20003-5, 20004-3 Two volumes, Paperbound $6.00

GREAT IDEAS OF OPERATIONS RESEARCH, Jagjit Singh. Easily followed non-technical explanation of mathematical tools, aims, results: statistics, linear programming, game theory, queueing theory, Monte Carlo simulation, etc. Uses only elementary mathematics. Many case studies, several analyzed in detail. Clarity, breadth make this excellent for specialist in another field who wishes background. 41 figures. x + 228pp.
21886-4 Paperbound $2.50

GREAT IDEAS OF MODERN MATHEMATICS: THEIR NATURE AND USE, Jagjit Singh. Internationally famous expositor, winner of Unesco's Kalinga Award for science popularization explains verbally such topics as differential equations, matrices, groups, sets, transformations, mathematical logic and other important modern mathematics, as well as use in physics, astrophysics, and similar fields. Superb exposition for layman, scientist in other areas. viii + 312pp.
20587-8 Paperbound $2.50

GREAT IDEAS IN INFORMATION THEORY, LANGUAGE AND CYBERNETICS, Jagjit Singh. The analog and digital computers, how they work, how they are like and unlike the human brain, the men who developed them, their future applications, computer terminology. An essential book for today, even for readers with little math. Some mathematical demonstrations included for more advanced readers. 118 figures. Tables. ix + 338pp.
21694-2 Paperbound $2.50

CHANCE, LUCK AND STATISTICS, Horace C. Levinson. Non-mathematical presentation of fundamentals of probability theory and science of statistics and their applications. Games of chance, betting odds, misuse of statistics, normal and skew distributions, birth rates, stock speculation, insurance. Enlarged edition. Formerly "The Science of Chance." xiii + 357pp.
21007-3 Paperbound $2.50

PLANETS, STARS AND GALAXIES: DESCRIPTIVE ASTRONOMY FOR BEGINNERS, A. E. Fanning. Comprehensive introductory survey of astronomy: the sun, solar system, stars, galaxies, universe, cosmology; up-to-date, including quasars, radio stars, etc. Preface by Prof. Donald Menzel. 24pp. of photographs. 189pp. 5¼ x 8¼.
21680-2 Paperbound $1.50

TEACH YOURSELF CALCULUS, P. Abbott. With a good background in algebra and trig, you can teach yourself calculus with this book. Simple, straightforward introduction to functions of all kinds, integration, differentiation, series, etc. "Students who are beginning to study calculus method will derive great help from this book." Faraday House Journal. 308pp.
20683-1 Clothbound $2.00

TEACH YOURSELF TRIGONOMETRY, P. Abbott. Geometrical foundations, indices and logarithms, ratios, angles, circular measure, etc. are presented in this sound, easy-to-use text. Excellent for the beginner or as a brush up, this text carries the student through the solution of triangles. 204pp.
20682-3 Clothbound $2.00

TEACH YOURSELF ANATOMY, David LeVay. Accurate, inclusive, profusely illustrated account of structure, skeleton, abdomen, muscles, nervous system, glands, brain, reproductive organs, evolution. "Quite the best and most readable account,' *Medical Officer.* 12 color plates. 164 figures. 311pp. 4¾ x 7.
21651-9 Clothbound $2.50

TEACH YOURSELF PHYSIOLOGY, David LeVay. Anatomical, biochemical bases; digestive, nervous, endocrine systems; metabolism; respiration; muscle; excretion; temperature control; reproduction. "Good elementary exposition," *The Lancet.* 6 color plates. 44 illustrations. 208pp. 4¼ x 7.
21658-6 Clothbound $2.50

THE FRIENDLY STARS, Martha Evans Martin. Classic has taught naked-eye observation of stars, planets to hundreds of thousands, still not surpassed for charm, lucidity, adequacy. Completely updated by Professor Donald H. Menzel, Harvard Observatory. 25 illustrations. 16 x 30 chart. x + 147pp.
21099-5 Paperbound $1.25

MUSIC OF THE SPHERES: THE MATERIAL UNIVERSE FROM ATOM TO QUASAR, SIMPLY EXPLAINED, Guy Murchie. Extremely broad, brilliantly written popular account begins with the solar system and reaches to dividing line between matter and nonmatter; latest understandings presented with exceptional clarity. Volume One: Planets, stars, galaxies, cosmology, geology, celestial mechanics, latest astronomical discoveries; Volume Two: Matter, atoms, waves, radiation, relativity, chemical action, heat, nuclear energy, quantum theory, music, light, color, probability, antimatter, antigravity, and similar topics. 319 figures. 1967 (second) edition. Total of xx + 644pp.
21809-0, 21810-4 Two volumes, Paperbound $5.00

OLD-TIME SCHOOLS AND SCHOOL BOOKS, Clifton Johnson. Illustrations and rhymes from early primers, abundant quotations from early textbooks, many anecdotes of school life enliven this study of elementary schools from Puritans to middle 19th century. Introduction by Carl Withers. 234 illustrations. xxxiii + 381pp.
21031-6 Paperbound $2.50

MATHEMATICAL PUZZLES FOR BEGINNERS AND ENTHUSIASTS, Geoffrey Mott-Smith. 189 puzzles from easy to difficult—involving arithmetic, logic, algebra, properties of digits, probability, etc.—for enjoyment and mental stimulus. Explanation of mathematical principles behind the puzzles. 135 illustrations. viii + 248pp.
20198-8 Paperbound $1.75

PAPER FOLDING FOR BEGINNERS, William D. Murray and Francis J. Rigney. Easiest book on the market, clearest instructions on making interesting, beautiful origami. Sail boats, cups, roosters, frogs that move legs, bonbon boxes, standing birds, etc. 40 projects; more than 275 diagrams and photographs. 94pp.
20713-7 Paperbound $1.00

TRICKS AND GAMES ON THE POOL TABLE, Fred Herrmann. 79 tricks and games— some solitaires, some for two or more players, some competitive games—to entertain you between formal games. Mystifying shots and throws, unusual caroms, tricks involving such props as cork, coins, a hat, etc. Formerly *Fun on the Pool Table*. 77 figures. 95pp.
21814-7 Paperbound $1.00

HAND SHADOWS TO BE THROWN UPON THE WALL: A SERIES OF NOVEL AND AMUSING FIGURES FORMED BY THE HAND, Henry Bursill. Delightful picturebook from great-grandfather's day shows how to make 18 different hand shadows: a bird that flies, duck that quacks, dog that wags his tail, camel, goose, deer, boy, turtle, etc. Only book of its sort. vi + 33pp. 6½ x 9¼. 21779-5 Paperbound $1.00

WHITTLING AND WOODCARVING, E. J. Tangerman. 18th printing of best book on market. "If you can cut a potato you can carve" toys and puzzles, chains, chessmen, caricatures, masks, frames, woodcut blocks, surface patterns, much more. Information on tools, woods, techniques. Also goes into serious wood sculpture from Middle Ages to present, East and West. 464 photos, figures. x + 293pp.
20965-2 Paperbound $2.00

HISTORY OF PHILOSOPHY, Julián Marias. Possibly the clearest, most easily followed, best planned, most useful one-volume history of philosophy on the market; neither skimpy nor overfull. Full details on system of every major philosopher and dozens of less important thinkers from pre-Socratics up to Existentialism and later. Strong on many European figures usually omitted. Has gone through dozens of editions in Europe. 1966 edition, translated by Stanley Appelbaum and Clarence Strowbridge. xviii + 505pp. 21739-6 Paperbound $3.00

YOGA: A SCIENTIFIC EVALUATION, Kovoor T. Behanan. Scientific but non-technical study of physiological results of yoga exercises; done under auspices of Yale U. Relations to Indian thought, to psychoanalysis, etc. 16 photos. xxiii + 270pp.
20505-3 Paperbound $2.50

Prices subject to change without notice.
Available at your book dealer or write for free catalogue to Dept. GI, Dover Publications, Inc., 180 Varick St., N. Y., N. Y. 10014. Dover publishes more than 150 books each year on science, elementary and advanced mathematics, biology, music, art, literary history, social sciences and other areas.